Function Transformations

Tim Brown
Naches Valley High School, Naches, WA, USA

Saltire Software, Inc.
Tigard, OR, USA
www.saltire.com
www.geometryexpressions.com

Saltire Software
P.O. Box 230755
Tigard, OR 97281-0755
http://www.geometryexpressions.com/
http://www.saltire.com/
support@saltire.com

Table of Contents

INTRODUCTION **5**

UNIT1: INTRODUCTION TO TRIGONOMETRY..**7**

   Lesson 1: Right Triangle Trigonometry 9

      Learning Objectives 9

      Overview for the Teacher.. 10

      Student Worksheets 10

   Lesson 2: The Unit Circle 13

      Learning Objectives 13

      Overview for the Teacher.. 14

      Student Worksheets 15

UNIT2: FUNCTION TRANSFORMATIONS **23**

   Lesson 1: Vertical Translations of Functions....................................... 27

      Learning Objectives 27

      Overview for the Teacher.. 28

      Student Worksheets 33

   Lesson 2: Vertical Dilations 43

      Learning Objectives 43

      Overview for the Teacher.. 44

      Student Worksheets 49

   Lesson 3: Combined Vertical Transformations 61

      Learning Objectives 61

      Overview for the Teacher.. 62

      Student Worksheets 69

   Lesson 4: Circular and Harmonic Motion... 83

      Overview for the Teacher.. 84

      Student Worksheets 89

   Lesson 5: Horizontal and Combined Transformations............................. 99

      Learning Objectives 99

      Overview for the Teacher.. 100

      Student Worksheets 112

   Lesson 6: Sinusoidal Curves 125

Learning Objectives 125
Overview for the Teacher 126
Student Worksheets 140

Extension A: Circles and Ellipses 157
Learning Objectives 157
Overview for the Teacher 158
Student Worksheets 160

Extension B: Absolute Value 165
Learning Objectives 165
Overview for the Teacher 166
Student Worksheets 169

Extension C: Cosine and Tangent 173
Learning Objectives 173
Overview for the Teacher 174
Student Worksheets 178

Extension D: Vertical Asymptotes 185
Learning Objectives 185
Overview for the Teacher 186
Student Worksheets 192

Introduction

The National Council of Teachers of Mathematics (NCTM) Vision for School Mathematics invites us to *"Imagine a classroom, a school, or a school district where all students have access to high-quality, engaging mathematics instruction."* It goes on to describe how this may take place in the classroom:

> *Teachers help students make, refine, and explore conjectures on the basis of evidence and use a variety of reasoning and proof techniques to confirm or disprove those conjectures. Students are flexible and resourceful problem solvers. Alone or in groups and with access to technology, they work productively and reflectively, with the skilled guidance of their teachers. Orally and in writing, students communicate their ideas and results effectively. They value mathematics and engage actively in learning it.* **(National Council of Teachers of Mathematics (NCTM).** *Principals and Standards for School Mathematics.* **Reston, VA: NCTM, 2000).**

Our goal in writing this book is to provide examples of how a symbolic geometry system, Geometry Expressions, can begin to make this happen. Geometry Expressions provides a playground where students can discover their own mathematics. They will begin to see mathematics as something that is created, not just a set of facts made up long ago. Once students take ownership of their mathematics, they will be more apt to "work productively and reflectively, with the skilled guidance of their teachers."

The graphical, interactive nature of Geometry Expressions brings life into a field that might otherwise seem irrelevant. The symbolics embedded in Geometry Expressions offer an algebraic view of the mathematics in concert with a geometric view, blurring the artificial line between the two. The smooth interface between Geometry Expressions and Computer Algebra Systems (CAS) adds another powerful resource for solving problems. These technologies can work together to change the way mathematics is done, in the same way that technology has changed the way architectural design is done; with computers managing the details while humans create the grand vision.

The units presented in this book are a jumping-off point for using Geometry Expressions in the classroom. Use the units to gauge the potential of this powerful software, and as a guide to applying Geometry Expressions in your own classroom. We trust that you will enjoy using the units and the software.

Unit1: Introduction to Trigonometry

Primary Mathematical Goals

- Students will transfer their knowledge of similar triangles to the definitions of the trigonometric functions.

- Students will expand the trigonometric functions to become the circular functions

Overview

The trigonometric functions are a natural bridge between the worlds of geometry and algebra (worlds that are much closer together than we are generally led to believe). Thus, it is a natural place to introduce the use of a symbolic geometry system.

The graphical, interactive nature of Geometry Expressions is ideal for the introduction of these concepts. Animation can be used to scale triangles while maintaining the values of their trig functions. A point can be rotated about the unit circle while illustrating the sign of the circular functions in each quadrant.

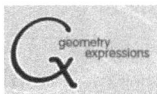

Lesson 1: Right Triangle Trigonometry

Learning Objectives

The student should develop an understanding of what trigonometric ratios essentially are through this lesson. It builds step-by-step from prior knowledge about similar triangles, through discovering equivalent ratios, to the formal definitions of sine, cosine, and tangent.

Math Objectives

- The student will learn the basic right-triangle definitions for sine, cosine, and tangent.

- The student will understand the connections between trigonometric ratios and their prior knowledge about similar triangles.

Technology Objectives

- The student will construct right triangles with given acute angle measures in Geometry Expressions.

Math Prerequisites

- The student should understand what makes similar triangles, and the proportionality of their sides.

Technology Prerequisites

- None. A basic understanding of Geometry Expressions helps, but is not required.

Materials

- A computer with Geometry Expressions for each student or pair of students.

- A scientific calculator.

Overview for the Teacher

This is an introduction, not a full lesson. It allows students to build an understanding of sine, cosine, and tangent as simply names for specific ratios of sides of similar right triangles. While this is a fairly common exercise to do with straightedge and protractor, the precision of student constructions is often not adequate for them to see the patterns. There could be a great deal of variation in the computed value for sine of a specific angle measure, for example. Using Geometry Expressions allows for the necessary precision.

1) All the triangles should be the same shape, but different sizes – i.e. they should be similar.

2) Student answers will vary, but rows 4, 5, & 6 should each have the same answer in all four columns. These should correspond to the sine, cosine, and tangent values of the student's chosen angle measure.

3) Students should get the same answer all the way across each row.

4) Angle-Angle Similarity theorem. This is a good small group discussion question. Have students review the theorem, and possibly the other similarity theorems while they're at it. If they haven't yet learned all the similarity theorems, this provides a sneak-preview of the topic, and you can discuss similarity in more general terms.

5) Answers will vary.

Students will need some guided and independent practice with the vocabulary at this point. Since such exercises are in virtually every Algebra 2 textbook, they are not reproduced here. Some suggested exercises would include:

- Given a right triangle and a specified angle, students name sides as opposite, adjacent, and hypotenuse. Use a variety of orientations.

- Given right triangles and all three sides, students produce sine, cosine, and tangent ratios for both acute angles.

- Students use a calculator to determine sine, cosine, and tangent ratios for given angles.

- Given a right triangle, all three sides, and a ratio, students determine if it is sine, cosine, or tangent for a given angle.

Student Worksheets

Student worksheets follow.

Name: _____

Date: _____

Right Triangle Trigonometry Intro

Pick a number between 10 and 80. Write it down here. _____

In Geometry Expressions, select File/New

Edit/Preferences/Math/Math/Angle Mode – set to degrees.

You are going to create a right triangle with one angle measure equal to the number you chose.

- Use **draw line segment** or **draw polygon** to create triangle ABC.

- Select angle C (Shift and click on AC, then BC), and use **constrain perpendicular** .

- Select angle A, and use **constrain angle** ; set your angle measure equal to the number you chose at the beginning.

- Highlight AB, then **calculate real distance** . This should give you a $Z_0 \Rightarrow$ and a number. Repeat with line segment BC and AC, which will give Z_1 and Z_2. These are the lengths of the line segments.

1) Now drag on any vertex of the triangle. What you notice about the size and shape of the triangle?

2) Complete the following table for four different versions of the triangle you constructed. They should all have the angle measure you chose at the beginning. When you get to the ratios, round to the thousandths place.

		Triangle #1	Triangle #2	Triangle #3	Triangle #4
1	Length AB				
2	Length AC				
3	Length BC				
4	$\dfrac{BC}{AB}$				
5	$\dfrac{AC}{AB}$				
6	$\dfrac{BC}{AC}$				

3) What do you notice about the rows 4-6?

4) Why is it safe to say that all triangles ABC with a right angle at C and your chosen angle measure at A will be similar to each other (i.e. each is a scale model of all the others)?

Since all right triangles with a given angle measure will be similar to each other, the ratios of their sides will be equal. This is what you saw in the last three rows. All of trigonometry is based on these facts about similar triangles and equal ratios. Because there are many, many applications and uses of the ratios of the sides of right triangles, mathematicians have given them specific names.

- Row 4 is called **Sine**: the ratio of the side opposite the angle to the hypotenuse.

- Row 5 is called **Cosine**: the ratio of the side adjacent to the angle to the hypotenuse.

- Row 6 is called **Tangent**: the ratio of the side opposite the angle to the adjacent side.

5) Use your calculator to determine the sine, cosine, and tangent of the angle measure you chose. Compare to the values in your table.

Sine:

Cosine:

Tangent:

A common memory trick to use is SOH-CAH-TOA.

$$\text{Sine} = \frac{Opposite}{Hypotenuse} \qquad \text{SOH}$$

$$\text{Cosine} = \frac{Adjacent}{Hypotenuse} \qquad \text{CAH}$$

$$\text{Tangent} = \frac{Opposite}{Adjacent} \qquad \text{TOA}$$

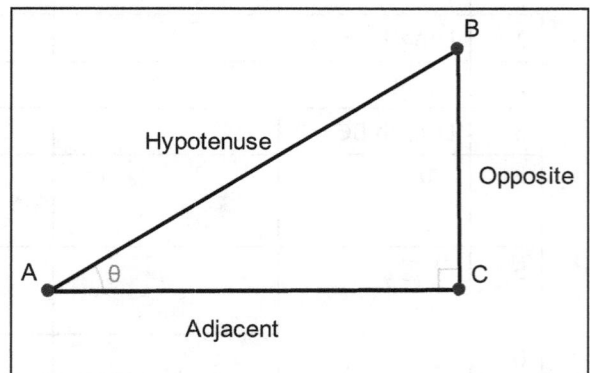

Lesson 2: The Unit Circle

Learning Objectives

This lesson is designed to transition students from an understanding of right triangle trigonometry to an understanding of the unit circle and trig ratios for angles greater than $90°$ or less than $0°$. They should develop a solid conceptual connection between right triangles and the unit circle, as well as understand the expanded definitions for sine, cosine, and tangent.

Math Objectives

- The student will understand the connections between the right triangle trigonometry and coordinates of a point on a unit circle.

- The student will understand and be able to use the expanded definitions of sine, cosine, and tangent for angles greater than $90°$ or less than $0°$.

Technology Objectives

- The student will become familiar with use of the various tools in the Geometry Expressions program, including drawing, constraining, animating, and constructing a locus

Math Prerequisites

- The student must understand and be proficient in right-triangle trigonometry.

- The student must understand a four-quadrant Cartesian coordinate system.

Technology Prerequisites

- Students should be able to complete this lesson with little to no background with Geometry Expressions.

Materials

- Computer with Geometry Expressions for each student or pair of students.

Overview for the Teacher

In this lesson, students will transition from an understanding of trigonometry that is limited to right triangles, to an understanding of the unit circle. This is done in a way that strongly emphasizes the logical connection and partial equivalence of the two aspects of trigonometry. Most students should be able to work through this lesson independently – it is not designed to require any direct instruction from the teacher. However, some common issues to look for and be aware of are outlined below.

This lesson involves a minimal amount of writing; most of the work will be done on the computer. You may want to institute a set of checkpoints, where you confirm the work has been done on the screen, and initial student papers. We suggest steps 8, 14, and 16 would be appropriate. Furthermore, the independent practice is not included, since exercises of this type are readily available. Following this lesson, students will need to do some practice (in worksheet form or otherwise) with...

- identifying coordinates of points on a unit circle given an angle of rotation
- identifying the sign of sine and cosine in the four quadrants
- correlating coordinates of points on a unit circle to their respective trig values.

Steps 1-5: In early versions of GX, some constraint contradictions tended to happen if one constructed the triangle by constraining point C to the axis directly or through its coordinates. If the location of C is established using a slope of 0 or direction of $0°$, the problem is avoided. This shouldn't be a problem in later versions of the software.

By step 8, student diagrams should resemble the following. (Numbers will be different depending on the angle measure chosen and the length of c.)

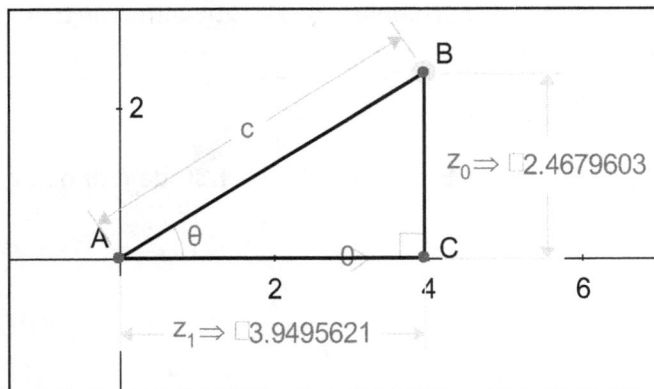

Step 14: Students may want to delete the side length measurements after they figure out the equivalence with the coordinates of point B. This is fine, and cleans up the diagram.

Step 16: The default variable for constructing a locus is c, but θ is needed. If students get a straight line instead of a circle, have them make sure they chose variable θ.

Final diagram should look like:

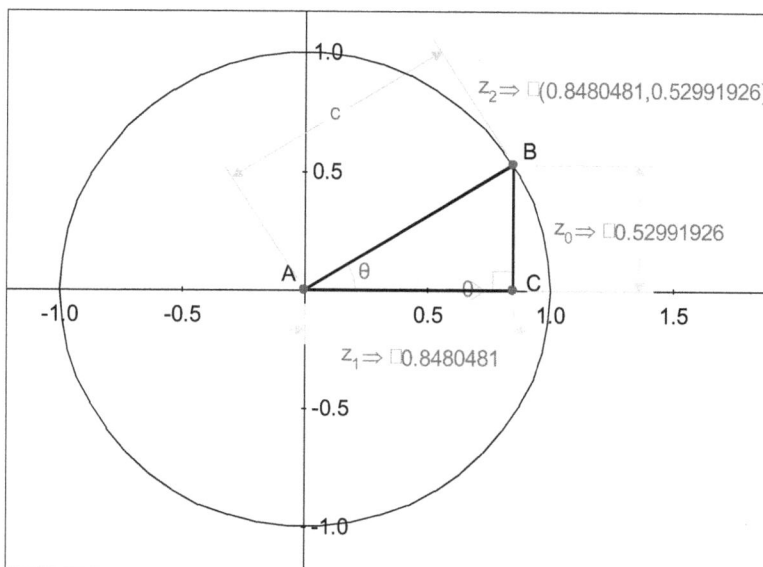

Student Worksheets

Student worksheets follow.

Name:_____

Date: _____

Intro to Unit Circle Trigonometry

Before you get started:

Open a new Geometry Expressions file. Make sure you are in degree mode and display your axes. (Edit/Preferences/Math/Math/Angle Mode -> Degrees; toggle the axis/grid button on the top toolbar ▦)

Right Triangle:

1) Pick a number between 10 and 80. Write it here: _____

2) You are going to create a triangle in the first quadrant. Use **draw line segment** to create triangle ABC, with A snapped to the origin, and B and C both somewhere in the first quadrant.

3) Highlight line segment AC and use **constrain slope** to set the slope to zero. This makes the segment line up with the x-axis.

4) Highlight line segments BC and AC at the same time by holding down the shift key and clicking on each. Use **constrain perpendicular** to form a right angle.

5) Highlight line segments AB and AC at the same time and use **constrain angle** to constrain the angle measure to θ.

You have just created a right triangle in **standard position**. A right triangle in standard position has the vertex with the angle you are concerned about (θ) at the origin, one leg of the triangle on the positive x-axis, and the third vertex in the first quadrant. Note that any right triangle could be positioned this way by rotating, translating, and/or reflecting it.

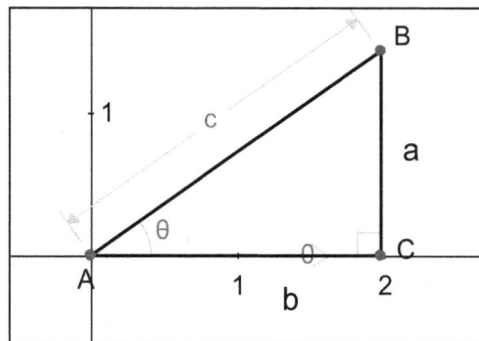

6) Highlight segment AB and use **constrain distance** to determine its length to be the variable c.

Variables		⁊ ✕
Variables	Functions	

Name	Value	Locke
c	4.4958507	-
θ	32	+

| θ | 32 | |

7) Now make the triangle your own. Use the Variables tool panel, highlight θ, then below that box, type the number you chose in part 1.

Lock θ by clicking on the icon next to the number.

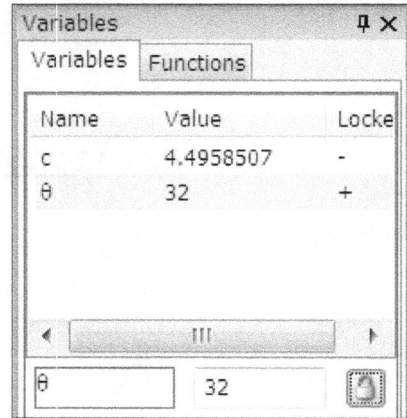

8) Use **calculate real distance** to calculate lengths a and b of your triangle; length c is displayed in the Variables window. Use those lengths to determine your sine, cosine, and tangent ratios.

Side	Length
a	
b	
c	

	Ratio	Decimal
Sine (θ)	/	
Cosine (θ)	/	
Tangent (θ)	/	

9) Use the trig functions on your calculator to check the ratios. Make sure your calculator is in degree mode.

10) What should happen to the triangle if we increase or decrease length c? Test your idea on GX by highlighting c in the Variables tool panel and changing the value. You can increase or decrease c by either typing in new numbers, or using the scroll bar underneath.

11) What will happen to the trig ratios as you do this?

12) For the rest of the activity, set *c* equal to 1 and lock it in the Variables tool panel.

You will probably need to scale up (button on the top tool bar – like zooming in) so the diagram takes up much of your screen.

13) Highlight point B and use **calculate real coordinates** to find the location of B. Compare this to your results in step 8. What do you notice?

14) Now adjust your diagram for some other angle measures by unlocking θ (click on the **lock/unlock** icon again), then either using the Variables tool panel, or clicking and dragging point B in your diagram. Test whether the trig ratios for two different angle measures correspond to coordinates for point B.

θ	Coordinates of B	Cos(θ)	Sin(θ)
	(,)		
	(,)		

Why does this work? If a right triangle is in standard position, the opposite side length is equal to the y-coordinate of point B, and the adjacent side length is equal to the x-coordinate of B.

Also, notice that we scaled the triangles back so that the hypotenuse is always 1. This can be done to any right triangle without changing the ratios of the sides.

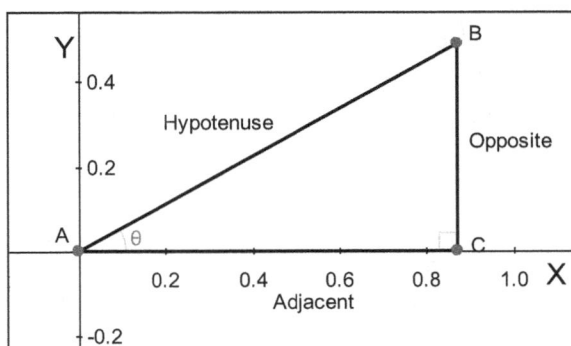

Therefore,

$$\sin\theta = \frac{opposite}{hypotenuse} = \frac{y}{1} = y$$

$$\cos\theta = \frac{adjacent}{hypotenuse} = \frac{x}{1} = x$$

$$\tan\theta = \frac{opposite}{adjacent} = \frac{y}{x} = \frac{Sin\theta}{Cos\theta}$$

These facts can be used to expand the definitions of sine, cosine, and tangent to angle measures greater than $90°$ and less than $0°$, as we will see shortly.

15) We want to look at all the possible values of θ. To do this, highlight θ in the Variables tool panel, and then look at the animation tools at the bottom of that section.

These tools allow you to move continuously through a range of values for a given variable. The bottom left and right boxes indicate the minimum and maximum values for θ. We want to see between $0°$ and $360°$. You can manually run through those values by clicking and dragging the scroll bar, or have the computer do it automatically by clicking on the play button. Notice the path that point B takes.

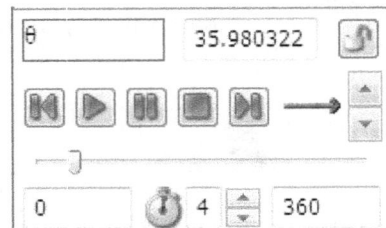

16) To see the pathway more clearly, highlight point B, and use the **construct locus** tool. You'll see a pop-up menu; make sure it is set to variable θ, start value 0, and end value 360. Click OK. Then repeat the animation from step 15.

When you consider the values of $0°$ through $90°$ - all the possibilities for a right triangle - the point B makes an arc. As you continue the pattern beyond the triangle, you create a circle with radius one, centered at the origin, which is called a **unit circle**. The coordinates of B, as we saw earlier, are $(\cos\theta, \sin\theta)$, where θ is the angle made with the positive x axis. The trick here is to think of θ a little differently. Instead of just being an acute angle of a right triangle, think of θ as the number of degrees that point B has rotated through since leaving the x-axis in the animation. For measures between $0°$ and $90°$, these two definitions of θ are exactly the same. However, the second definition allows us to use angle measures greater than $90°$.

17) Similarly, angle measure of less than $0°$ indicate a rotation in the opposite direction. Adjust your beginning value in the variables menu to -360, and use the scroll bar to investigate.

In short, the expanded definitions of sine and cosine are the coordinates of a point on a unit circle, after the point has rotated θ degrees counterclockwise from the positive x-axis. The coordinates are $(\cos\theta, \sin\theta)$. Similarly, tangent of θ becomes $\frac{y}{x}$.

18) Pause the animation at various angle measures in the different quadrants and determine from the coordinates where sine is positive and negative, and where cosine is positive and negative. Complete the diagram below with the words "positive," and "negative."

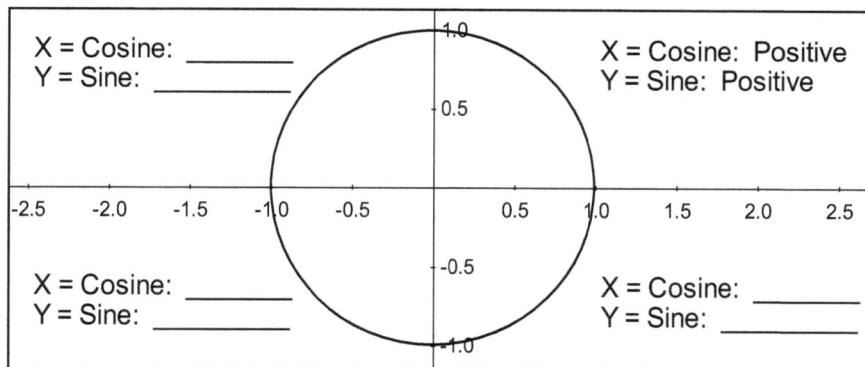

X = Cosine: _____   X = Cosine: Positive
Y = Sine: _____   Y = Sine: Positive

X = Cosine: _____   X = Cosine: _____
Y = Sine: _____   Y = Sine: _____

19) Based on what you have learned so far, or by observing the points on the unit circle, what is the greatest and least value you could get for sine of some angle θ? This is called the range of sine.

20) What is the greatest and least value you could get for cosine of some angle θ? This is the range of cosine.

Unit2: Function Transformations

Primary Mathematical Goals:

- Students will understand and be able to create vertical and horizontal translations and dilations of functions.

- Students will be able to determine and articulate what transformations have been done to a function based on an equation or a graph.

- Students will understand and be able to apply the connections between equations, graphs, and verbal/contextual problems.

- Students will apply the above skills and knowledge to three function families: quadratics, inverse variations, and sine curves.

Overview:

This unit is designed to familiarize students with the ideas of how various functions can be transformed, and the effect those transformations have on equations, graphs, and contextual situations. It specifically investigates three basic function types: those with parent functions $y = x^2$, $y = \dfrac{1}{x}$, and $y = \sin x$. These function families were chosen because they are simple enough for students to readily understand, and also sophisticated enough to clearly demonstrate the effects of dilations and translations. The skills and principles to be learned apply to virtually all function families, and will give a solid foundation for more advanced studies in functions.

The approach of this unit is primarily investigative in nature – students will examine the effects various transformations have on function equations and graphs to develop conjectures and generalized understanding. To that end, the Geometry Expressions (GX) software will be used extensively as an investigative tool and a means to check the accuracy of conjectures. Ideally, students will have previously completed the Intro to Unit Circle Trigonometry lesson using GX, as that introduces the general learning pattern, as well as many of the software features they will be using. If they haven't used GX before, it is a good lesson to do as a review, prior to starting this unit.

The unit starts with vertical dilations and translations, as those tend to be easier for students to understand. The first two lessons are abstract mathematics, with practical modeling coming in the third lesson: combining vertical translations and dilations. It is advisable to pose one or two of the questions from lesson three to students at the beginning of the unit, so they have the practical purpose of the work in mind as they go. Lesson four combines vertical translations and dilations in the sine curve, and the modeling problems in it tend to foreshadow the more realistic sinusoidal curve problems of lesson six. Horizontal dilations are only studied in the sine curve, since they can be clearly distinguished from vertical dilations in that function family. In the process of learning about function transformations, students also review area dilations, investigate the vertex form of quadratics, study basic asymptotes, and deepen their understanding of circular and harmonic motion.

Four "extension" lessons are also included. The extensions go beyond the core purpose of the unit, and do not have to be taken in order; the teacher can pick and choose lessons to include. The first of these lessons extends the transformation ideas to graphs of non-functions. Specifically, the relationships between circles and ellipses, and the derivations of their equations are explored. The second and third extension lessons relate the ideas to other function families, namely absolute value, cosine, and tangent. The final extension explores vertical asymptotes, and in the process introduces the secant, cosecant, and cotangent functions.

Outline

Lesson 1: Vertical Translations   *Time Required: 50 – 80 min.*

- Part A: Investigates vertical translations in three function families.

- Part B: Independent practice with vertical translations in other functions.

Lesson 2: Vertical Dilations   *Time Required: 50 – 80 min.*

- Part A: Investigates vertical dilations in three function families.

- Part B: Independent practice with vertical dilations in other functions.

Lesson 3: Combined Vertical Transformations/Applications   *Time Required: 90 – 120 min.*

- Part A: Investigates combined vertical transformations in quadratics via area models.

- Part B: Potentially independent work with combined vertical transformations in projectile models and applied inverse variation models.

Lesson 4: Circular and Harmonic Motion   *Time Required: 120 – 150 min.*

- Students apply vertical transformations to the sine curve to create equations, graphs, and simulations of simplified circular and harmonic motion.

Lesson 5: Horizontal and Combined Transformations   *Time Required: 200 - 250 min.*

- Part A: Investigates horizontal translations, and combines them with vertical transformations in quadratics. Investigates vertex form.

- Part B: Combines horizontal translations with vertical transformations in inverse variation functions – both abstract and applied models.

Lesson 6: Sinusoidal Curves   *Time Required: 200 – 250 min.*

- Horizontal and vertical translations and dilations are all combined in various circular and harmonic motion models. Students work with equations, graphs, and simulations.

Extension A: Circles and Ellipses   *Time Required: 35 – 50 min.*

- Students apply transformations to non-function graphs and derive equations.

Extension B: Absolute Value Functions   *Time Required: 20 – 55 min.*

- Students apply transformations to absolute value functions.

Extension C: Cosine and Tangent Functions   *Time Required: 50 – 75 min.*

- Students apply transformations to cosine and tangent functions. Students create circular and harmonic motion models with cosine.

Extension D: Vertical Asymptotes   *Time Required: 60 – 80 min.*

- Students explore vertical asymptotes in rational and reciprocal trigonometric functions.

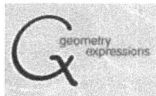

Lesson 1: Vertical Translations of Functions

Learning Objectives

This lesson gives a basic investigative approach to understanding the effect of adding a constant to a function.

Math Objectives

- The student will understand that the effect of adding a constant to a function is a vertical translation.

- The student will be able to determine the effect a change in the equation has on a graph, and vice verse with respect to vertical translations.

- The student will review the shapes of graphs and some key characteristics of the functions $y = x^2$, $y = \dfrac{1}{x}$, and $y = \sin(x)$

Technology Objectives

- The student will become proficient with the function feature of GX, and with using the variables tool to do dynamic investigations.

Math Prerequisites

- Students must know what a function is, and be familiar with three parent functions: $y = x^2$, $y = \dfrac{1}{x}$, and $y = \sin(x)$

- Students must be comfortable with radian measure for the sine function.

- Students should be familiar with function notation: $y = f(x)$.

Technology Prerequisites

- Students should have a basic familiarity with Geometry Expressions. This can be accomplished with the "Intro to Unit Circle Trigonometry" lesson, or chapter 2 of the professional development materials.

Materials

- A computer with Geometry Expressions for each student or pair of students.

- Colored pencils (optional, but recommended).

Overview for the Teacher

Throughout this unit, we will be looking at function transformations through analysis of three families of functions, whose parent functions are $y = x^2$, $y = \dfrac{1}{x}$, and $y = \sin(x)$. For today's lesson, students will investigate vertical translations graphically, and analyze the correlation with symbolically adding a constant to a function. Vertical translations tend to be simple for students, and so the lesson also reviews some characteristics of each function family and helps students become more familiar with the software at the same time. With the use of the interactive software, students should be able to deduce the main ideas fairly easily on their own. The teacher's role today is mostly to facilitate and keep students on track. Students may need clarification and/or assurance that they understand the primary pattern correctly. You will particularly want to confirm student understanding of asymptotes and their equations on question 3, before they move on.

The second part should be done without the help of GX, so you may want to copy it on a different paper and hand it out as students leave the computers to work on their own. The software is a great tool for developing the ideas, but if we allow it to be overused, it can become a means for students to hide their lack of understanding. It is important that they can use the concepts they develop independently, and it is equally important that they know they are expected to learn the math, not just the computer application. However, the software can and will be used far more extensively later in the unit, for dynamic modeling, etc.

It will be helpful for students to have colored pencils and color-code their different transformations of the graphs. This is highly recommended both for ease of student understanding, and for ease of checking/grading.

For some classes, this lesson may get overly repetitive. Teachers should modify/shorten as appropriate for their particular group of students.

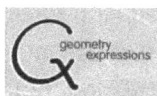

Vertical Translations part A

1) Teachers may want to introduce the animation tool at this point

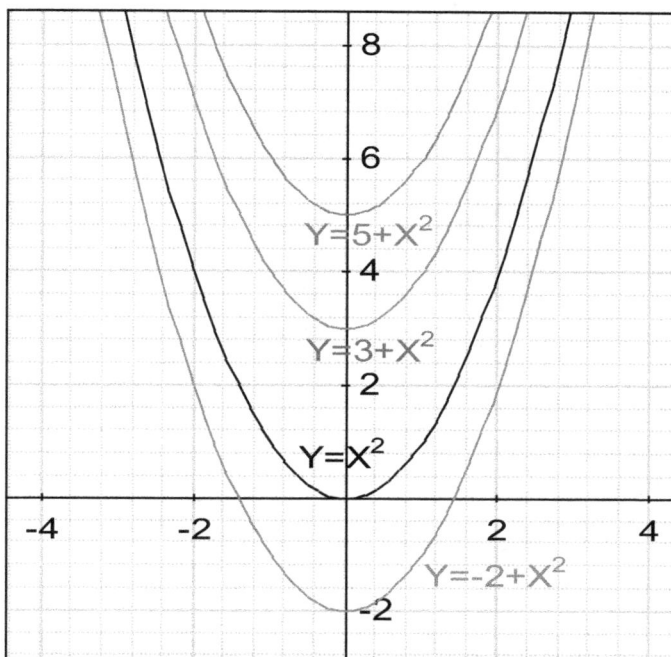

2) Answers may vary slightly. The graph is moving up (vertical translation) by the number of units that is being added to the function. If it is a negative number, the graph is moving down.

3) A)

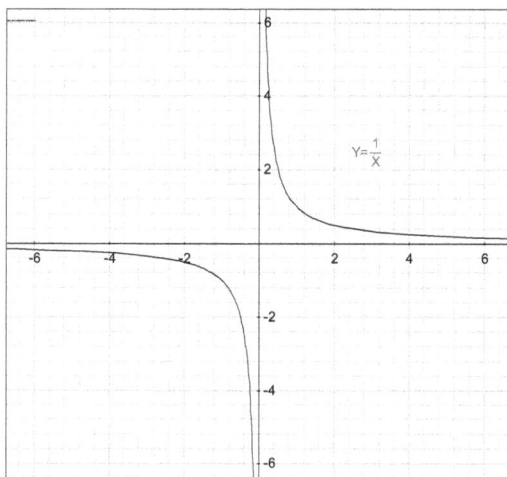

B) A vertical line through the origin;
A horizontal line through the origin.

C) Y = 0; X = 0

It is important to check student understanding of asymptotes and their equations before they move on.

Function Transformations Lesson 1 -Vertical Translations of Functions
Algebra 2; Pre-Calculus
Time required: 50 – 80 min.

4)
    A) Shifted the graph up 2 units (vertical translation).

    B) Y = 2
       X = 0

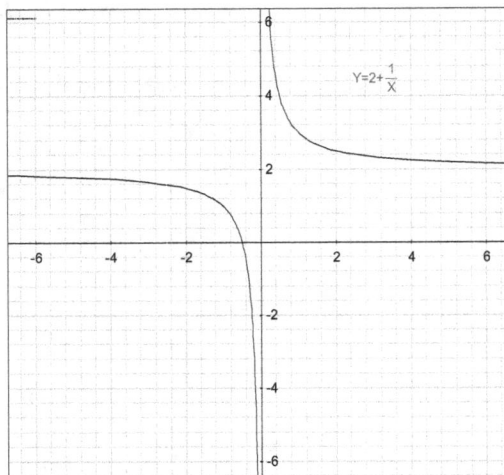

5)
    A) Shifted the graph up 4 units (vertical translation).

    B) Y = 4
       X = 0

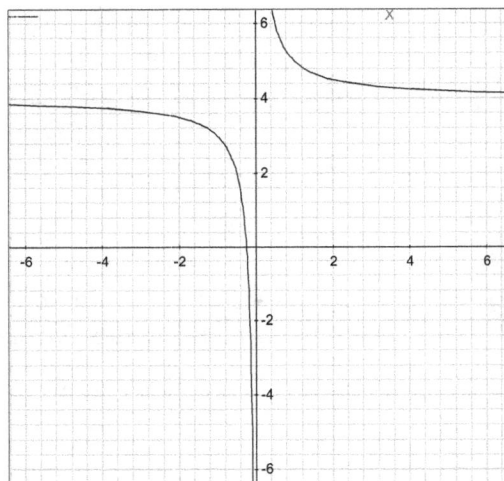

6)
    A) Shifted the graph down 3 units (vertical translation).

    B) Y = -3
       X = 0

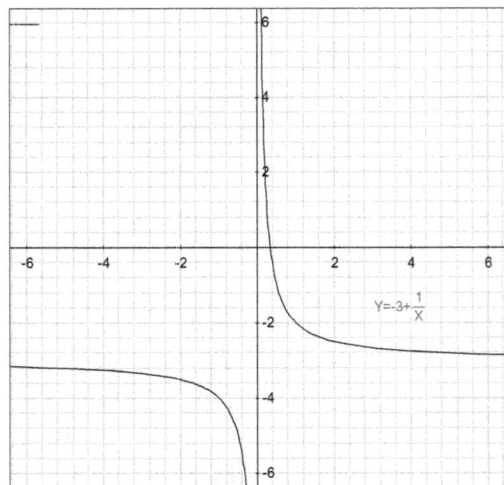

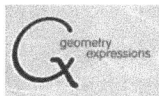

7)

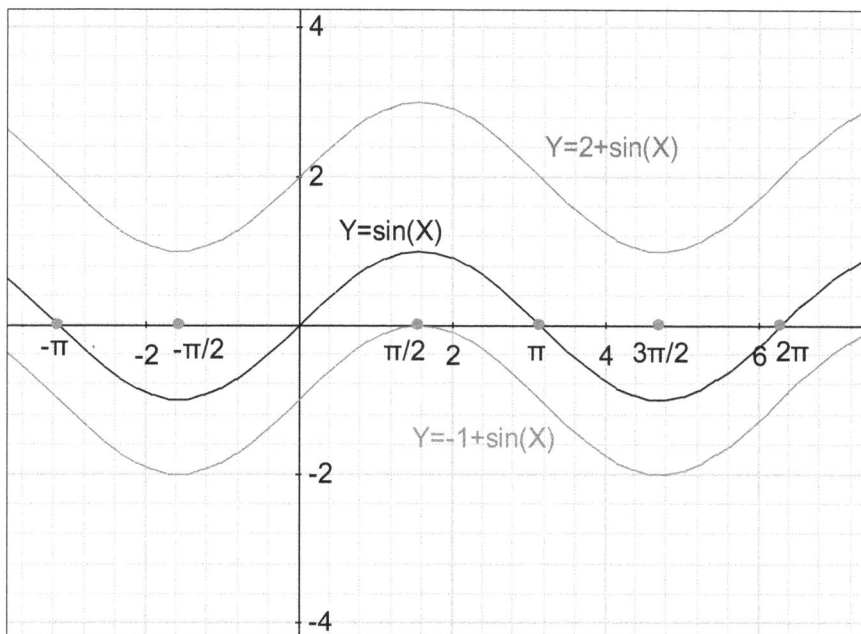

8) The number added to a function shifts (translates) the function vertically by that many units. A positive number slides the graph up; a negative number slides the graph down.

Teacher Modeling: As students transition from the computer (possibly group) work to the independent practice, you may want to model the vertical translation of an arbitrary function on the board. A common issue is that students will draw a very rough, sloppy approximation of the general shape. Model the fact that they need to look at specific points and add the translation amount to each y-value. You may particularly want to model a curve, giving the example of identifying relative maximum and minimum values as reference points.

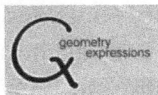

Vertical Translations Part B: Independent Practice

1-2)

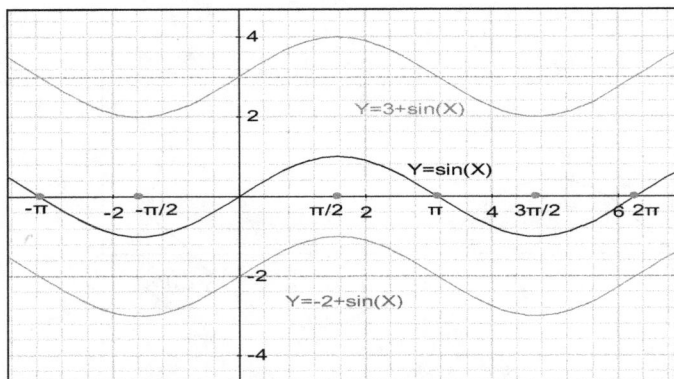

3-5)

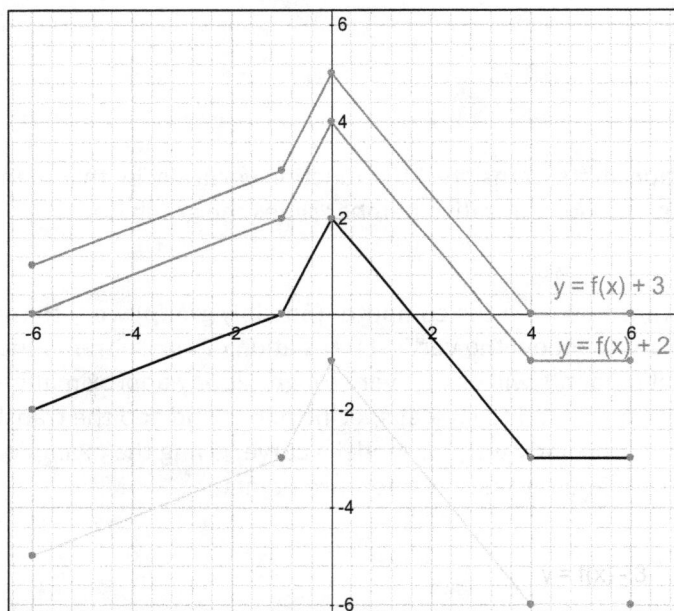

6-7)

Points labeled on graph: $(3, 6.4)$, $Y=3+g(X)$, $(3, 3.4)$, $(-2, 0.4)$, $Y=g(X)$, $(-2, -2.6)$, $(3, -2.6)$, $Y=-6+g(X)$, $(-2, -8.6)$

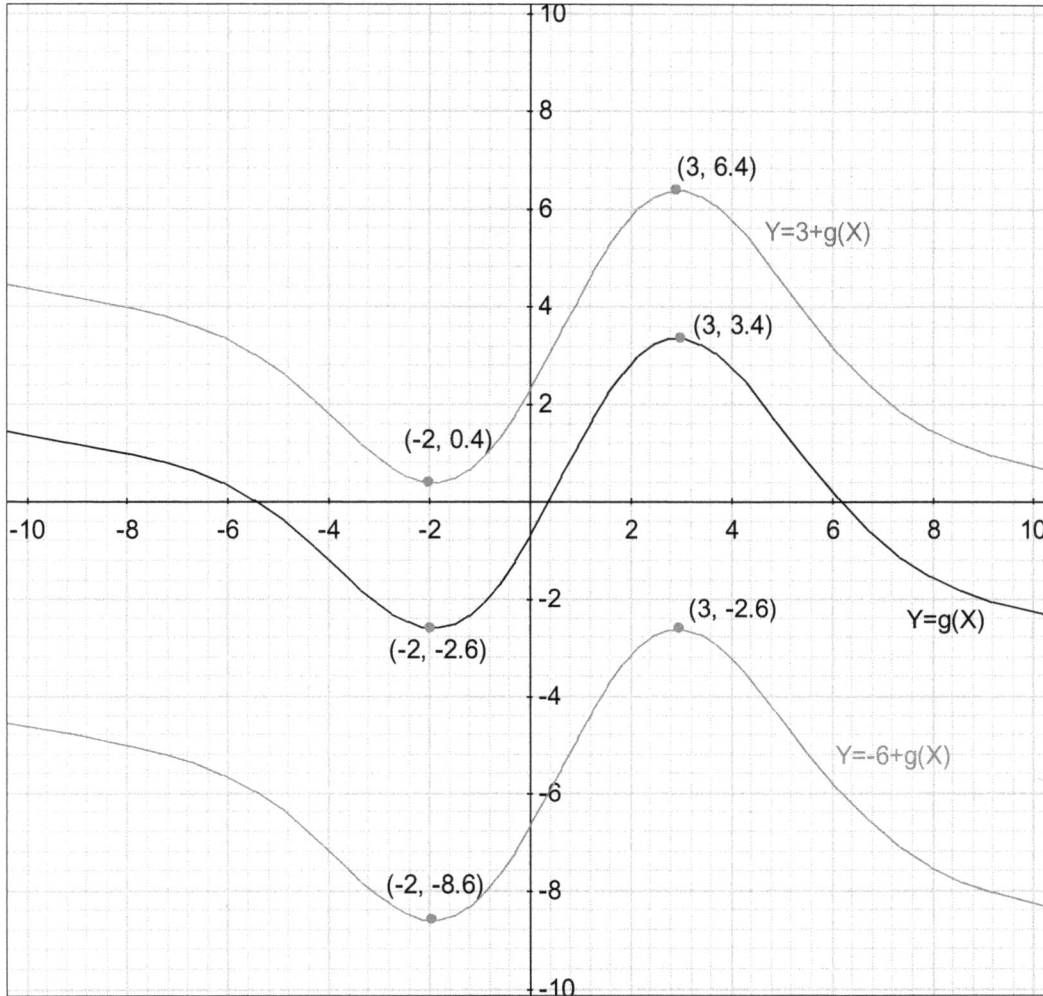

8) $j(x) = h(x) + \underline{3}$
      $h(x) = j(x) + \underline{-3}$

9) $m(x) = k(x) + 3$
      $k(x) = m(x) - 3$

Student Worksheets

Student worksheets follow.

Vertical Translations part A

I. $y = x^2$

The function $y = x^2$ is the parent for a family of functions, whose graphs are a shape called a parabola. The vertex, or local maximum or minimum, is a key point for a parabola.

1) Draw the graphs below and investigate the larger pattern of what happens to the function as you add a constant number to it, then answer the questions below. Clearly label and/or color-code your graphs.

   GX will help you if you use the **draw function** tool ⬚, and type in $y = x^2 + c$. Be careful, you must use * to indicate multiplication and ^ to indicate an exponent. Then use the variables tool panel, highlight c, and type in the appropriate number in the box. You can see how the graph changes for a wider range of a values by clicking on the curve and dragging up and down, or by highlighting c in the Variables tool panel and dragging the scroll bar.

   A) $y = x^2$ B) $y = x^2 + 3$

   C) $y = x^2 + 5$ D) $y = x^2 - 2$

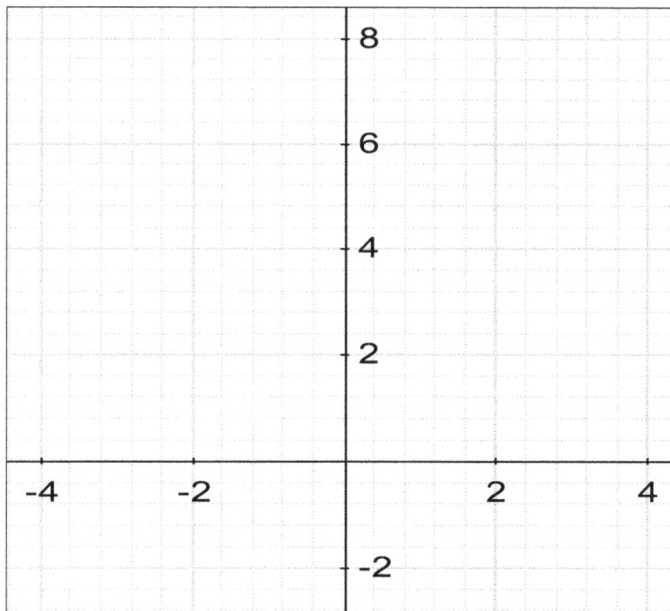

2) What seems to happen to the graph of the function when a constant number is added to it? Be sure to include the case where the number is negative.

II. $y = \dfrac{1}{x}$

The function $y = \dfrac{1}{x}$ is the parent of a family of functions whose graphs are a shape called a hyperbola. Each of these graphs will have two <u>asymptotes</u>. An asymptote is a line that the graph approaches, but never reaches.

1) A) Draw in a graph of $y = \dfrac{1}{x}$ below. Always use a straightedge and draw in a dotted line to indicate an asymptote.

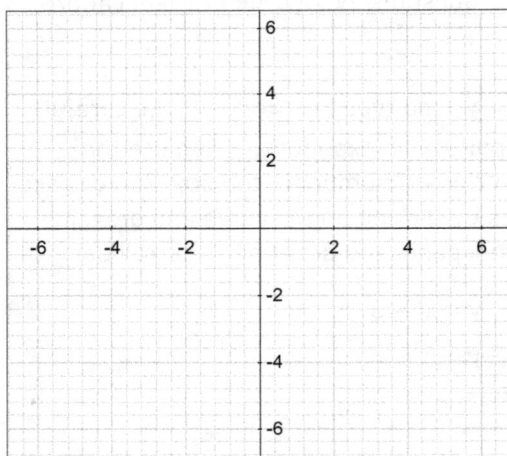

B) Describe the positions of the two asymptotes in words.

C) Write an equation for each of the asymptotes:

Y =

X =

Using GX, graph $y = \dfrac{1}{x} + c$. Assign different values for c to create graphs of the functions below. In each case, describe the effect the "c" value had on the graph, and write the equations for the asymptotes. Also select c in the Variables tool panel and drag the scroll bar to get a more complete idea of how different possible values affect the graph.

2) $y = \dfrac{1}{x} + 2$

A) Effect of the "c" value?

B) Equations for the asymptotes.

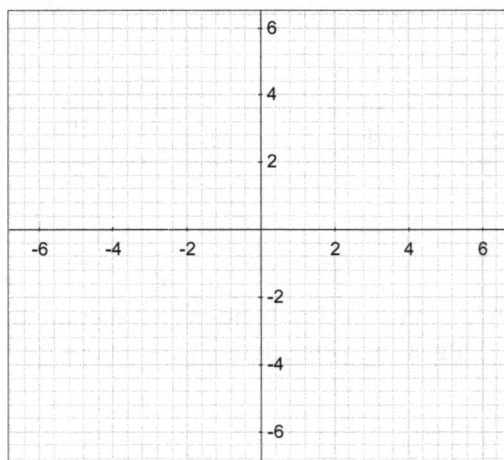

3) $y = \dfrac{1}{x} + 4$

    A) Effect of the "c" value?

    B) Equations for the asymptotes.

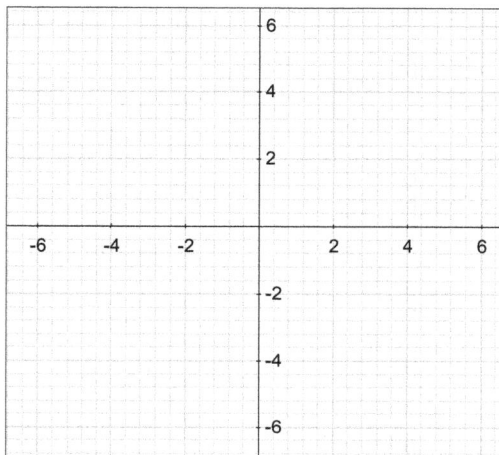

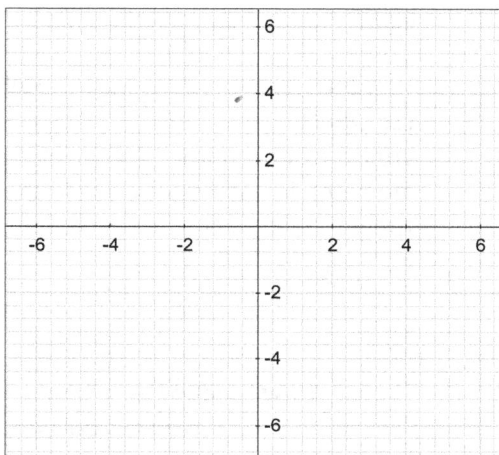

4) $y = \dfrac{1}{x} - 3$

    A) Effect of the "c" value?

    B) Equations for the asymptotes.

III. $y = Sin(x)$

The third family of functions we will be exploring in this unit has the parent function $y = \sin(x)$. Remember that this is a periodic function, which means the pattern repeats itself at regular intervals. For the sine function, the period is 2π, or approximately 6.28. Also remember that it has a maximum value of 1, and a minimum value of –1, and it oscillates across a central axis, which is the line y = 0 (the x-axis).

1) A)   Draw the graph of $y = \sin(x)$. This can be done from memory, or using GX.

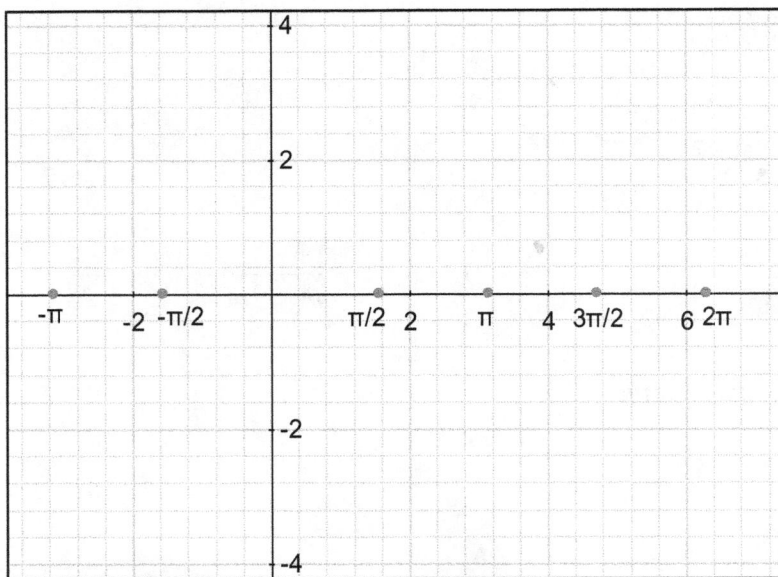

Using GX, graph $y = \sin(x) + c$. Make sure you set the program to radian mode (Edit/Settings/Math/Math/Angle Mode -> Radians). Assign different values for *c* to create graphs of the other given functions. (Your graphs on GX won't have π, 2π, etc. labeled.) In each case, draw in the central axis as a dotted line. Also use the scroll bar in the Variables tool panel again to explore the effect *c* has.

B) y = sin(x) +2

C) y = sin(x) − 1

2) On GX, use **draw function** to graph y = f(x). This is a generic, arbitrary function. Now graph y = f(x) + c. Use the variables tool window to vary the values of *c*. What seems to happen to the graph of any function when a constant number is added to it? Be sure to include the case where the number is negative.

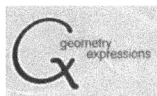

Vertical Translations Part B: Independent Practice

The following problems should be done without the help of a computer or graphing calculator. You are encouraged to use your responses to part A as a reference. On each problem, pay attention to scale and specific key points – don't just sketch in a vague shape.

1) The graph of y = sin(x) is given below. On the same axes, draw in a graph of y = sin(x) + 3.

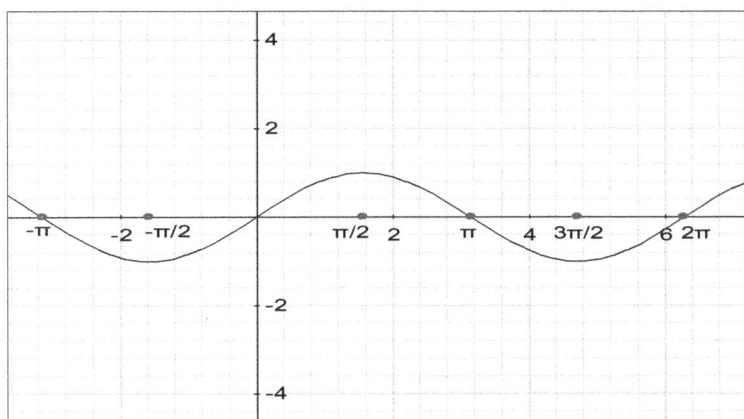

2) On the axes above, draw in a graph of y = sin(x) - 2. Label and/or color-code your graph.

3) Below is the graph of a function y = f(x). On the same set of axes, draw in a graph of y = f(x) + 2. Label and/or color-code your graph.

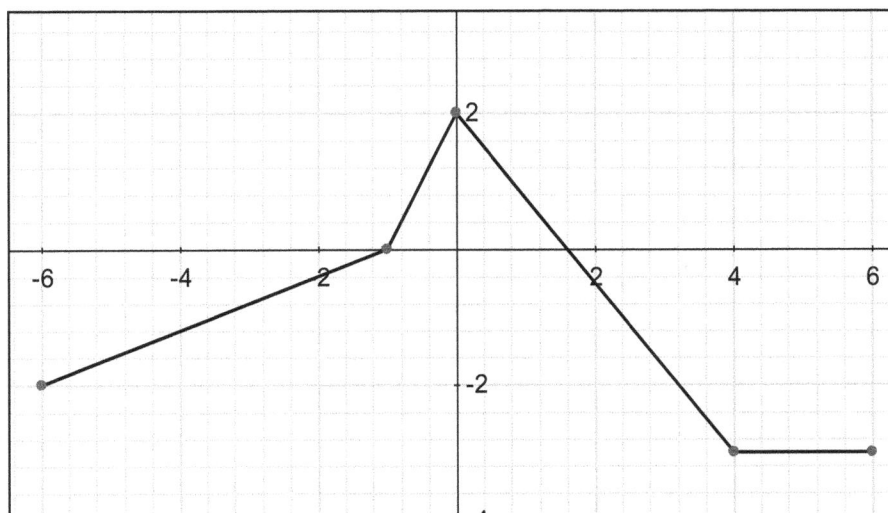

4) Now draw in a graph of y = f(x) + 3.

5) Now draw in a graph of the function y = f(x) - 3.

6) Below is the graph of a function y = g(x). On the same set of axes, draw in a graph of y = g(x) + 3.

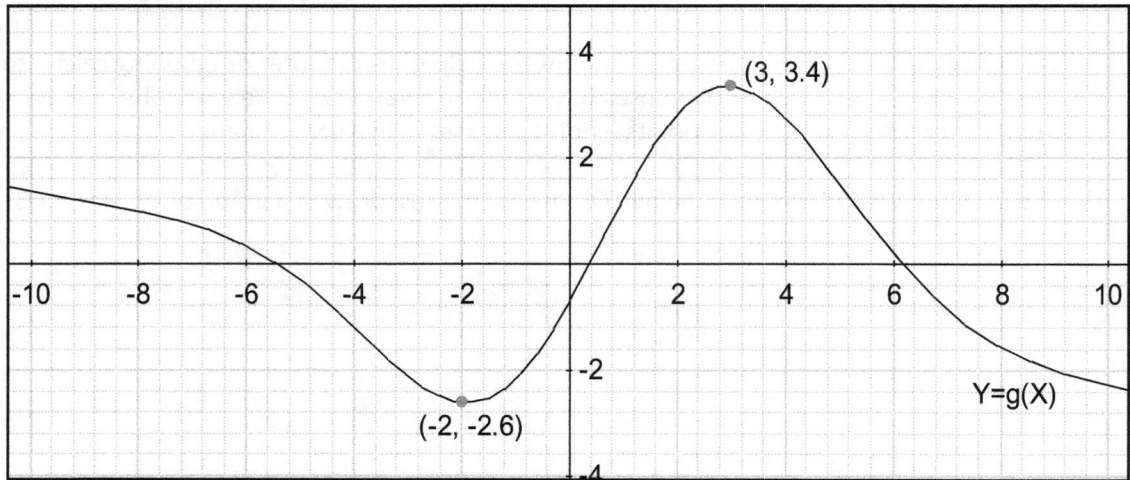

7) Now draw in a graph of the function y = g(x) - 6 above. Label and/or color-code your graphs.

8) Graphed below are two functions, h(x) and j(x), which have a simple relationship to each other. One is a vertical translation of the other. Write out that relationship two ways, by filling in the blanks of the following equations:

$$j(x) = h(x) + \underline{\hspace{2cm}}$$

$$h(x) = j(x) + \underline{\hspace{2cm}}$$

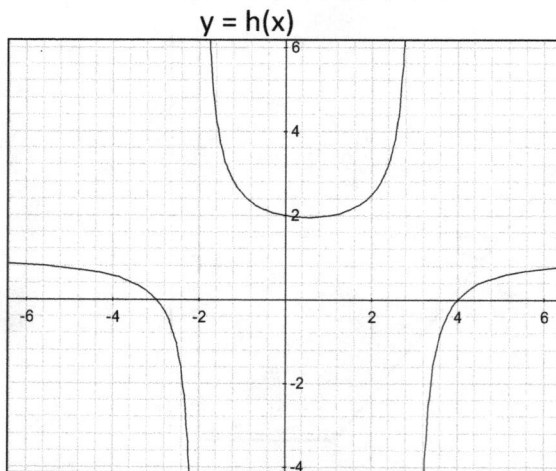

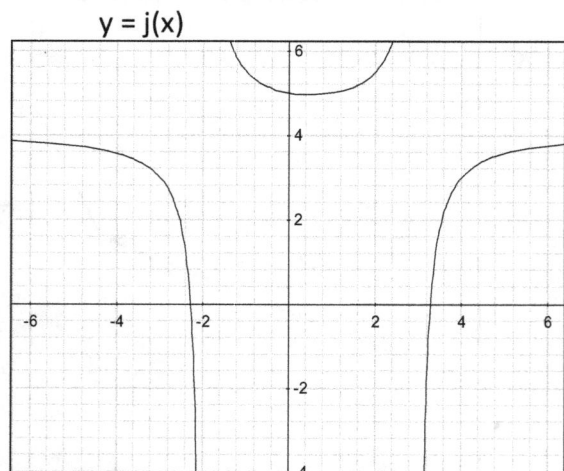

y = h(x)

y = j(x)

9) Graphed below are two functions, k(x) and m(x), which have a simple relationship to each other. Write out that relationship two ways, by writing two equations.

m(x) =

k(x) =

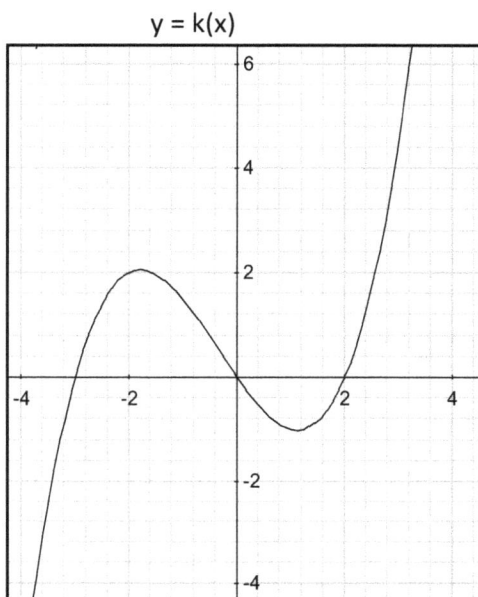

y = k(x)

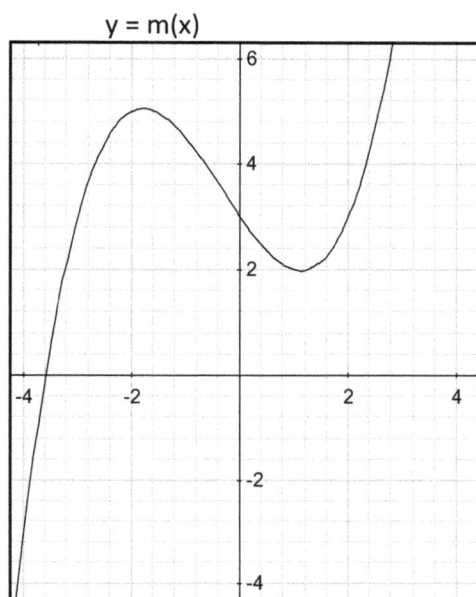

y = m(x)

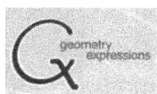

Lesson 2: Vertical Dilations

Learning Objectives

This lesson gives a basic investigative approach to understanding the effect of multiplying a function by a constant.

Math Objectives

- The student will understand that the effect of multiplying a function by a constant is a vertical dilation.

- The student will be able to determine the effect a change in the equation has on a graph, and vice verse with respect to vertical dilations.

Technology Objectives

- The student will become proficient with the function feature of GX, and with using the variables tool to do dynamic investigations.

Math Prerequisites

- Students must know what a function is, and be familiar with three parent functions: $y = x^2$, $y = \dfrac{1}{x}$, and $y = \sin(x)$

- Students must be comfortable with radian measure for the sine function.

- Students should be familiar with function notation: $y = f(x)$.

Technology Prerequisites

- Students should have a basic familiarity with Geometry Expressions.

Materials

- A computer with Geometry Expressions for each student or pair of students.

- Colored Pencils (optional, but recommended).

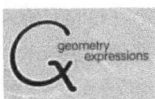

Overview for the Teacher

For today's lesson, students will investigate vertical dilations of the three function families graphically, and analyze the correlation with multiplying a function equation by a constant. With the use of the interactive software, students should be able to deduce the main ideas fairly easily on their own. The teacher's role today is mostly to facilitate and keep students on track. There is at least one key point where some up-front modeling will help as students transition to independent work. Students may need clarification and/or assurance that they understand the primary pattern correctly.

Part B should be done without the help of GX, so you may want to copy it on a different paper and hand it out as students leave the computers to work on their own.

It will be helpful for students to have colored pencils and color-code their different transformations of the graphs. This is highly recommended both for ease of student understanding, and for ease of checking/grading.

For some classes, this lesson may get overly repetitive. Teachers should modify/shorten as appropriate for their particular group of students.

Vertical Dilations Part A

1) Teachers may want to introduce the animation tool at this point.

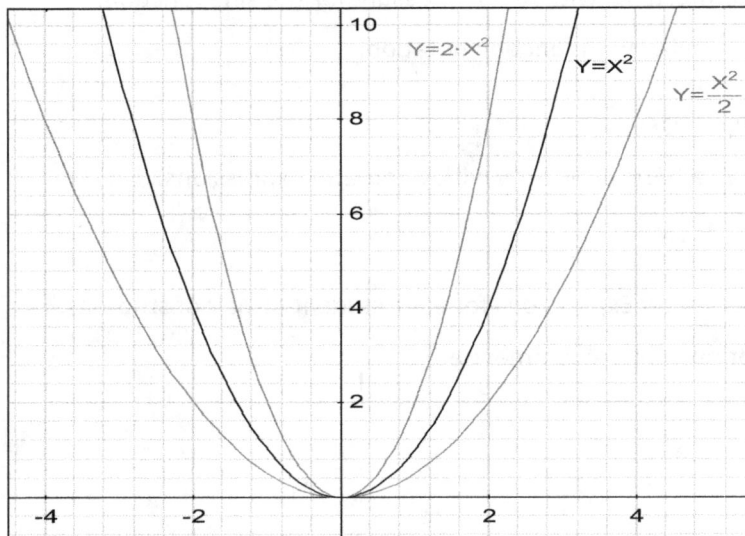

2) Answers will vary somewhat. As a increases, the graph gets stretched (dilated) vertically. Some students will describe this as a horizontal shrink instead, which is actually accurate for this particular function. Horizontal and vertical dilations become distinct when students examine the sine curve.

A good discussion to have with students while they are working is what happens when $a = 0$. They should be able to algebraically determine that this creates the function equation $y = 0$, which is coincident with the x-axis. They can very clearly see how this fits into the rest of the pattern by allowing a to vary from -2 to 2, and using the scroll bar, the animation feature, or simply dragging on a point on the curve. This can tie the specific answers to the questions together with a dynamic view of the overall patterns at work.

3)

A, B, C)

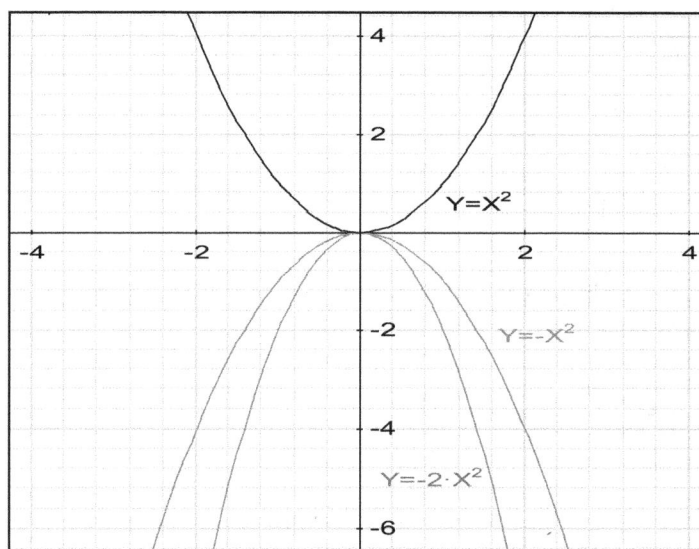

D) Answers will vary somewhat. Multiplying by negative one flips the graph vertically – reflection over the x-axis. Multiplying by other negative numbers yields the reflection and a vertical stretch (dilation).

4)

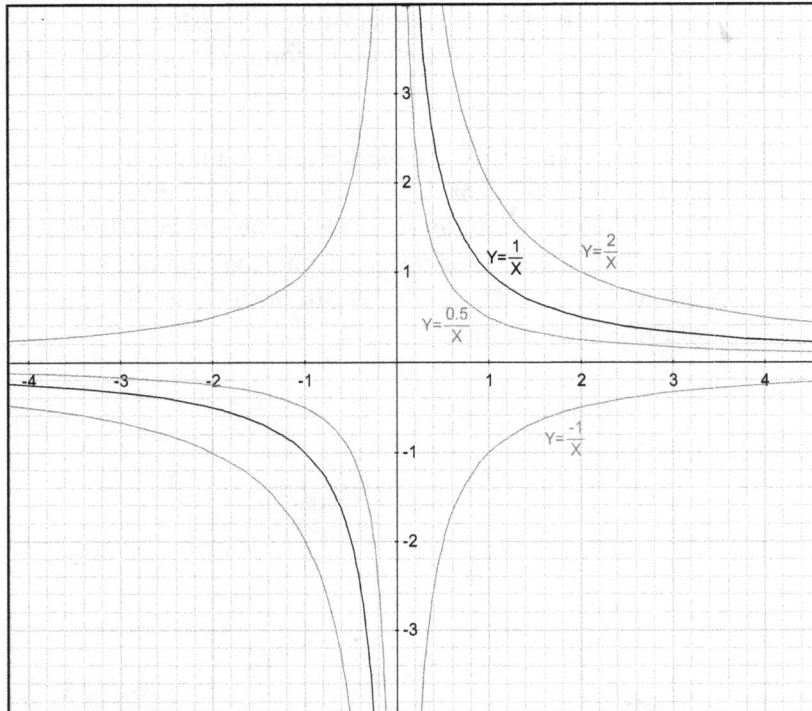

5) Answers will vary somewhat. Multiplying by a constant a should stretch the graph vertically. If a is negative, the graph is reflected across the x-axis and stretched.

6)

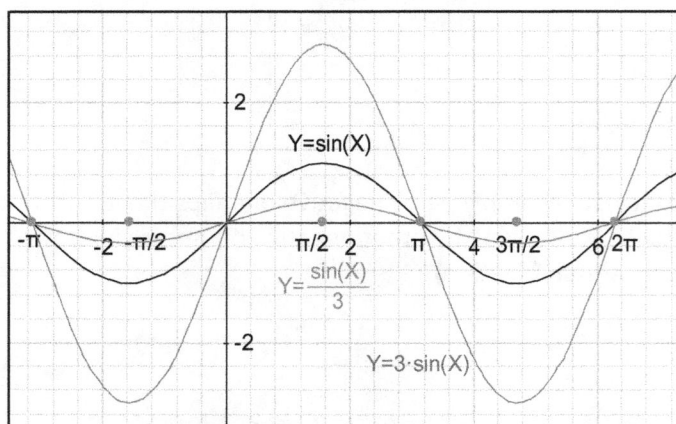

7) Multiplying a function by a stretches it vertically (vertical dilation) by a scale factor of a.

8) Answers will vary greatly. One possibility is given:

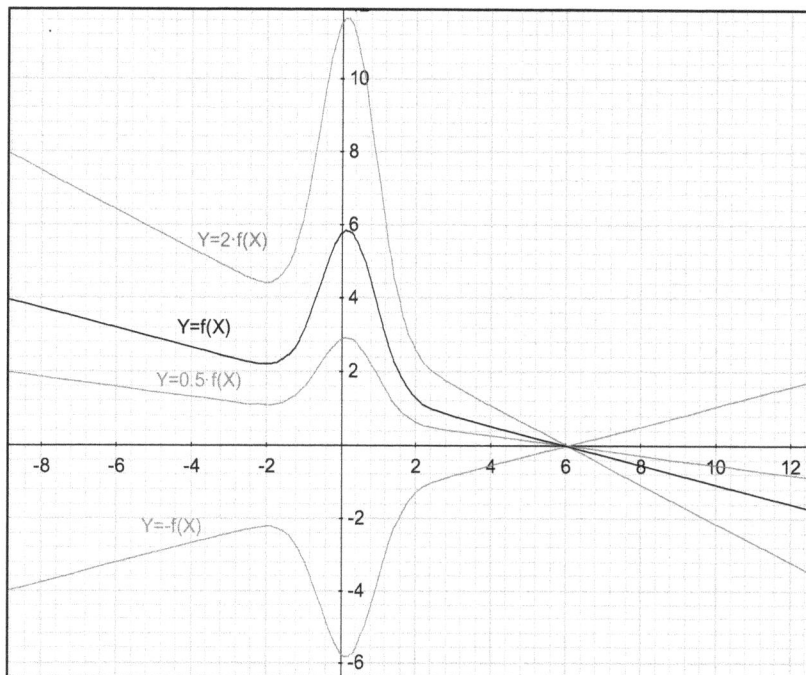

9)

A) Vertical stretch or vertical dilation by a scale factor of *a*.

B)

- If *a > 1*, the graph gets stretched "taller" vertically

- If *a < 0*, the graph gets flipped – reflected across the x axis – then stretched or shrunk

- If *0 < a < 1* the graph shrinks vertically – gets "shorter"

Teacher Modeling: As students transition from the computer (possibly group) work to the independent practice, you may want to model the vertical dilation of an arbitrary function on the board. A common issue is that students will draw a very rough, sloppy approximation of the general shape. Model the fact that they need to look at specific points and multiply the y-value of each by the scale factor given. You may particularly want to model a curve, giving the example of identifying relative maximum and minimum values to scale up as reference points. Also have students take note of the fact that any point on the x-axis stays there.

Vertical Dilations Part B: Independent Practice

1 & 2)

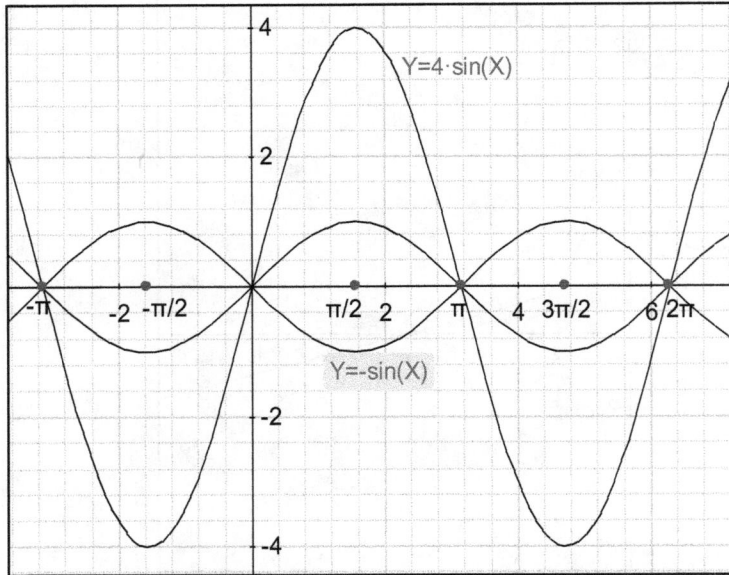

3 – 5)

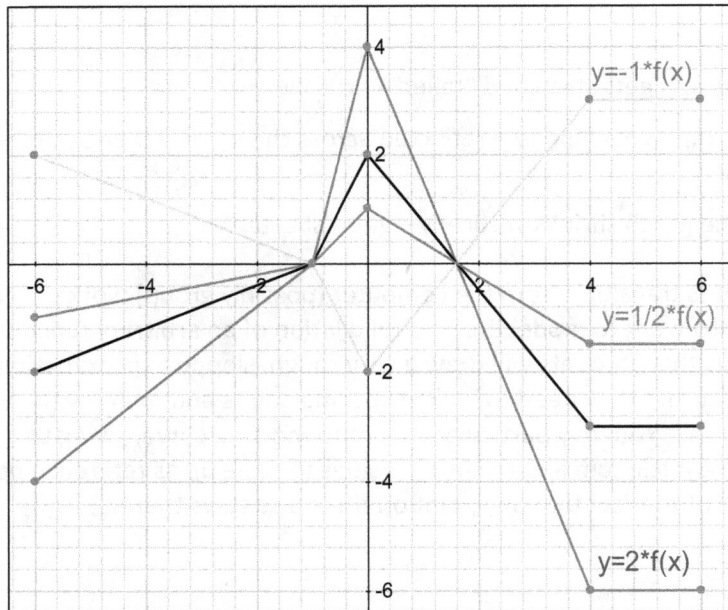

6 – 7)

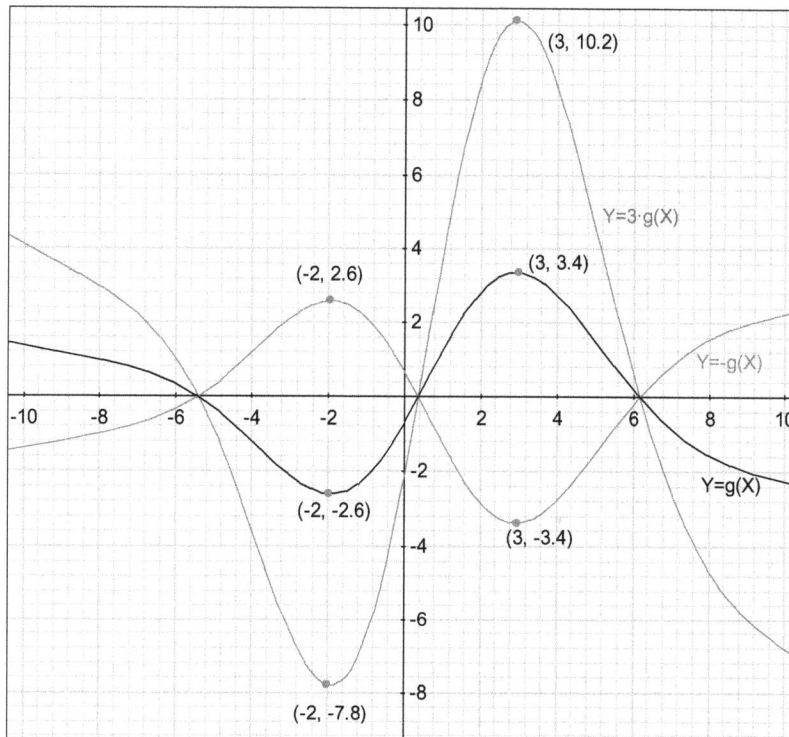

8)   h(x) = 1/2 * j(x)
       j(x) = 2* h(x)

9)   k(x) = 4 * m(x)
       m(x) = ¼ * k(x)

Student Worksheets

Student worksheets follow.

Name: _____

Date: _____

Vertical Dilations Part A

I. $y = x^2$

1) Graph the functions below. These functions all belong to the same family, and $y = x^2$ is the parent function. You can use GX to help you if you use the **draw function** tool , and type in the equation. Be careful, you must use * to indicate multiplication (even if a number is next to a variable), and ^ to indicate an exponent.

A) $y = x^2$

B) $y = 2x^2$

C) $y = \dfrac{1}{2}x^2$

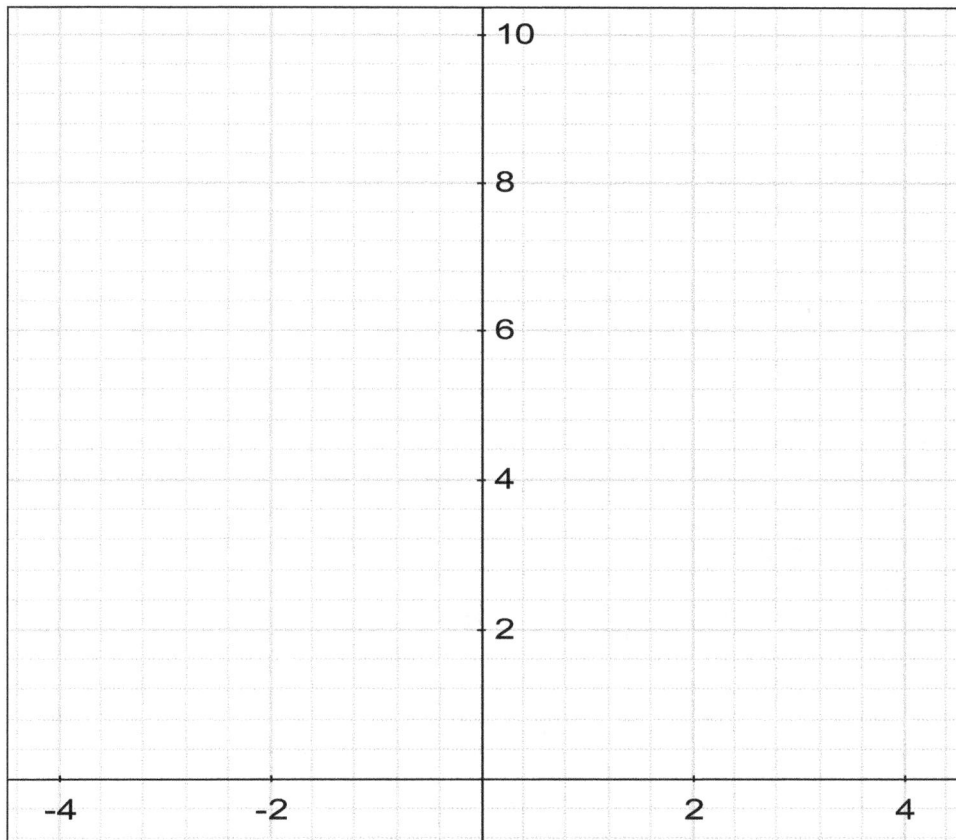

2) Now try to determine what happens as the coefficient (the number in front of the variable) changes. Use the function tool on GX to graph $y = ax^2$. You can see how the graph changes by clicking on the curve and dragging up and down, or by highlighting a in the Variables tool panel and dragging the scroll bar. Investigate values from at least 0 through 5. As a increases (i.e. the parent function is multiplied by greater number), what happens to the graph?

3) Graph the following two functions, and then use the graph of $y = ax^2$ to help you determine what happens to the function when you multiply it by a negative number.

      A) $y = x^2$

      B) $y = (-1)x^2$

      C) $y = (-2)x^2$

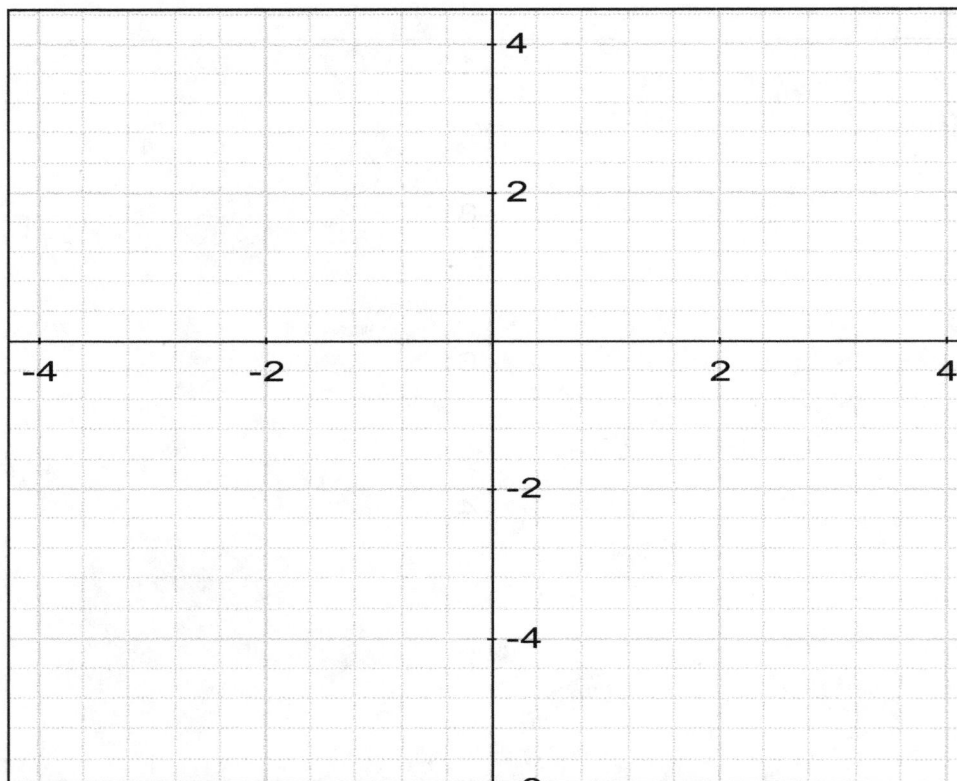

      D)   Use the scroll bar to investigate values from –5 through 5. What was the effect of multiplying the function by a negative number?

II. $y = \dfrac{1}{x}$

1) Now we're going to do a similar exercise with the family of functions related to $y = \dfrac{1}{x}$. Open a new GX file and graph $y = \dfrac{a}{x}$. Note that this is just any constant number a times the parent function. Assign different values for a to create graphs of the functions below. Color-code and/or label each function. Also, use the scroll bar in the Variables tool panel to get a more complete idea of the effect changing a is having.

A) $\quad y = \dfrac{1}{x}$

B) $\quad y = \dfrac{2}{x}$

C) $\quad y = \dfrac{0.5}{x}$

D) $\quad y = \dfrac{-1}{x}$

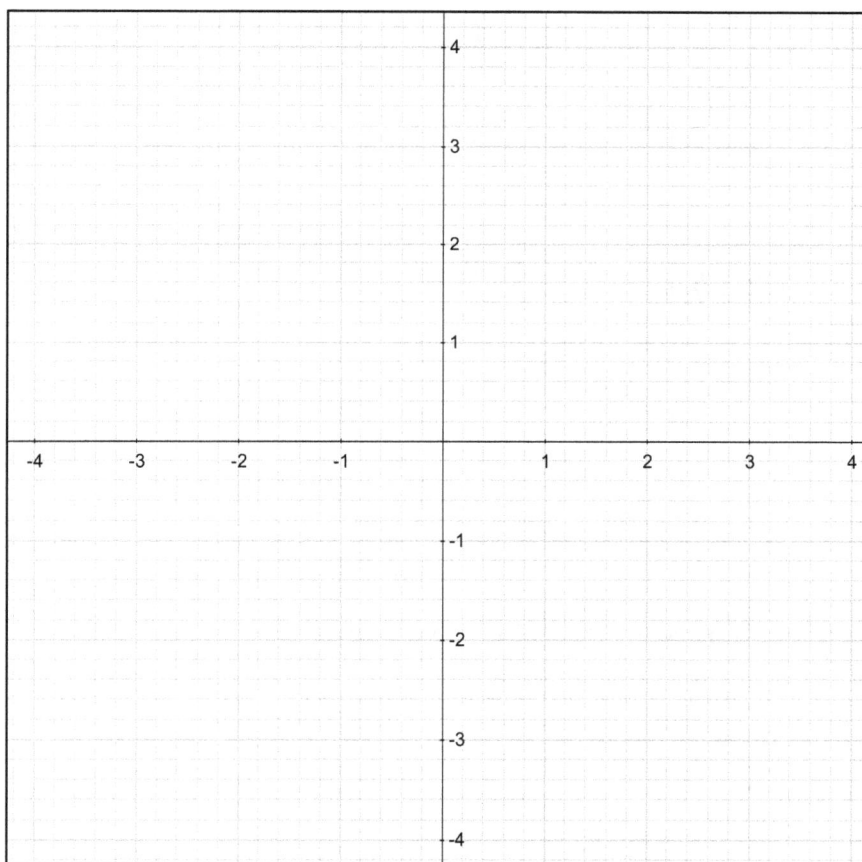

2) What effect did multiplying the parent function by a constant a have on the graph? What about negative values for a?

III. $y = Sin(x)$

1) Open another new GX file and graph $y = a*\sin(x)$. Make sure you set the program to radian mode (Edit/Settings/Math/Math/Angle Mode -> Radians). Assign different values for *a* to create graphs of the functions below. (Your GX graphs won't have π, 2π, etc. labeled.) Color-code and/or label each function. Also, use the scroll bar in the Variables tool panel to get a more complete idea of the effect changing *a* is having.

     A) $y = \sin(x)$

     B) $y = 3\sin(x)$

     C) $y = \dfrac{1}{3}\sin(x)$

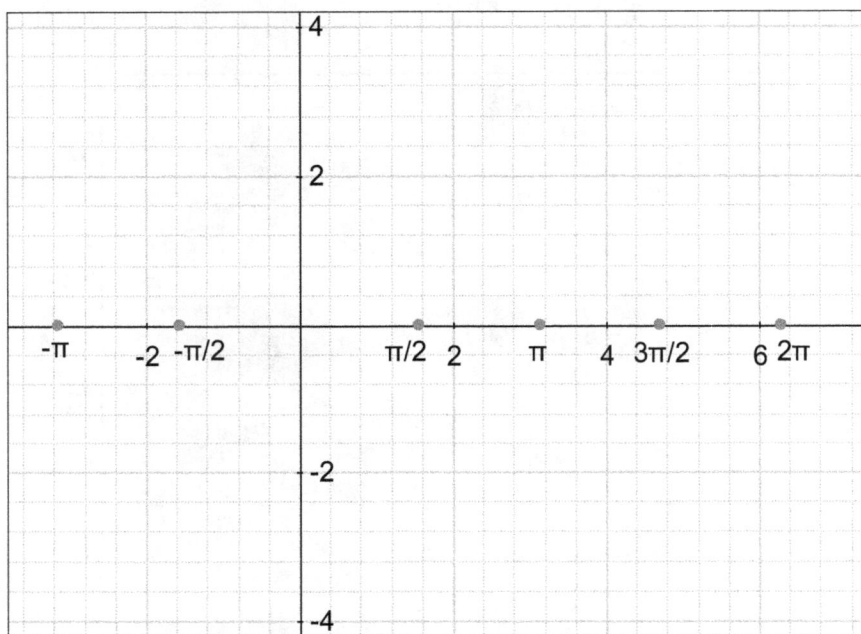

2) What effect does multiplying this function by a constant number have on its graph?

IV. Y = Generic Function:

1) We are now going to try, as nearly as we can, to examine the effect of multiplying any arbitrary function by a constant. Open a new GX file and draw the function y = f(x), y = g(x) or y = h(x). Click and drag the function around a few times so it looks different from your neighbor's. Now graph y = *a*f(x)* [or *a*g(x)*, etc]. Watch what happens as you vary the value of *a* with the scroll bar. Sketch and label four graphs below: your parent function and three different *a* values. Let one have *a > 1*, one have *a < 0*, and one have *0 < a < 1*.

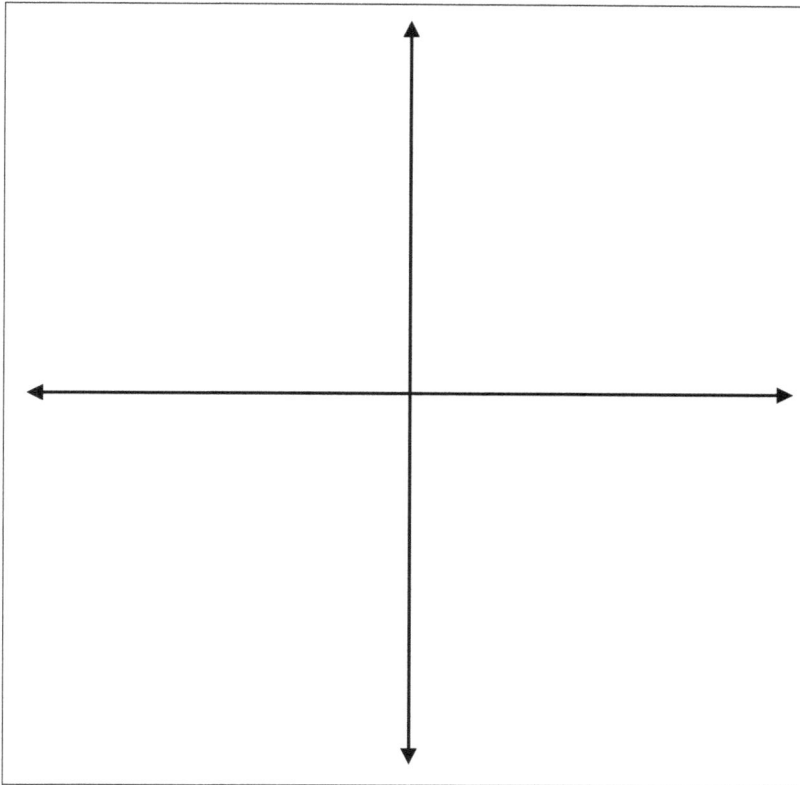

2) At this point, you should be convinced that multiplying a function by a constant will have the same effect on any function. If you are unsure of this, go back and re-examine your four sets of functions.

   A) Describe in general terms: What is the effect of multiplying a function by a constant *a*?

B) Be more specific about the effect for different values of *a*:

- *a > 1*

- *a < 0*

- *0 < a < 1*

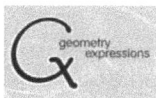

Name: _____

Date: _____

Vertical Dilations Part B: Independent Practice

The following problems should be done without the help of a computer or graphing calculator. You are encouraged to use your responses to part A as a reference. On each problem, pay attention to scale and specific key points – don't just sketch in a vague shape.

1) The graph of $y = sin(x)$ is given below. On the same axes, draw in a graph of $y = 4*sin(x)$.

2) On the axes below, draw in a graph of $y = (-1)*sin(x)$.

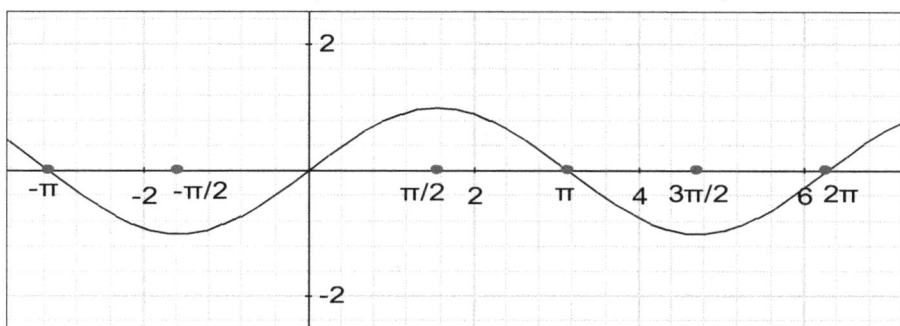

3) Below is the graph of a function $y = f(x)$. On the same set of axes, draw in a graph of $y = 2*f(x)$. Label and/or color-code your graph.

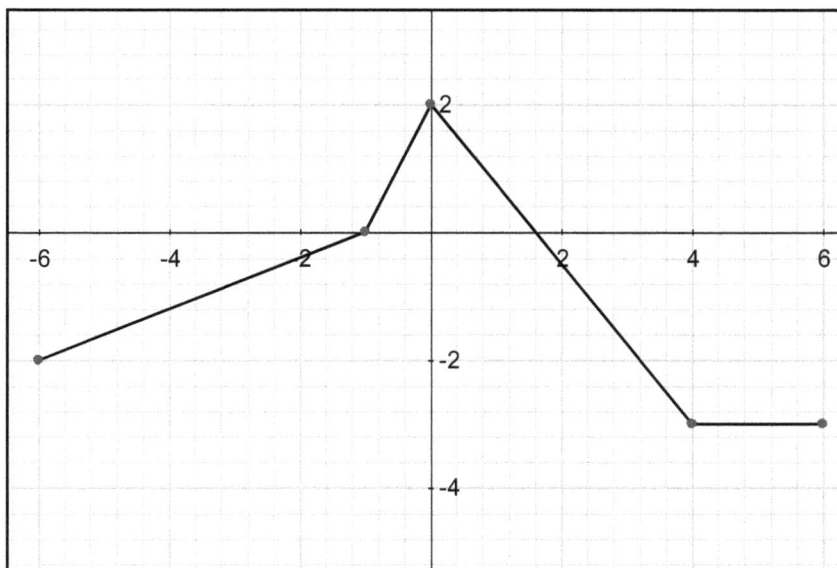

4) Now draw in a graph of $y = \frac{1}{2}*f(x)$.

5) Now draw in a graph of the function $y = -1*f(x)$.

6) Below is the graph of a function y = g(x). On the same set of axes, draw in a graph of y = 3 *g(x).

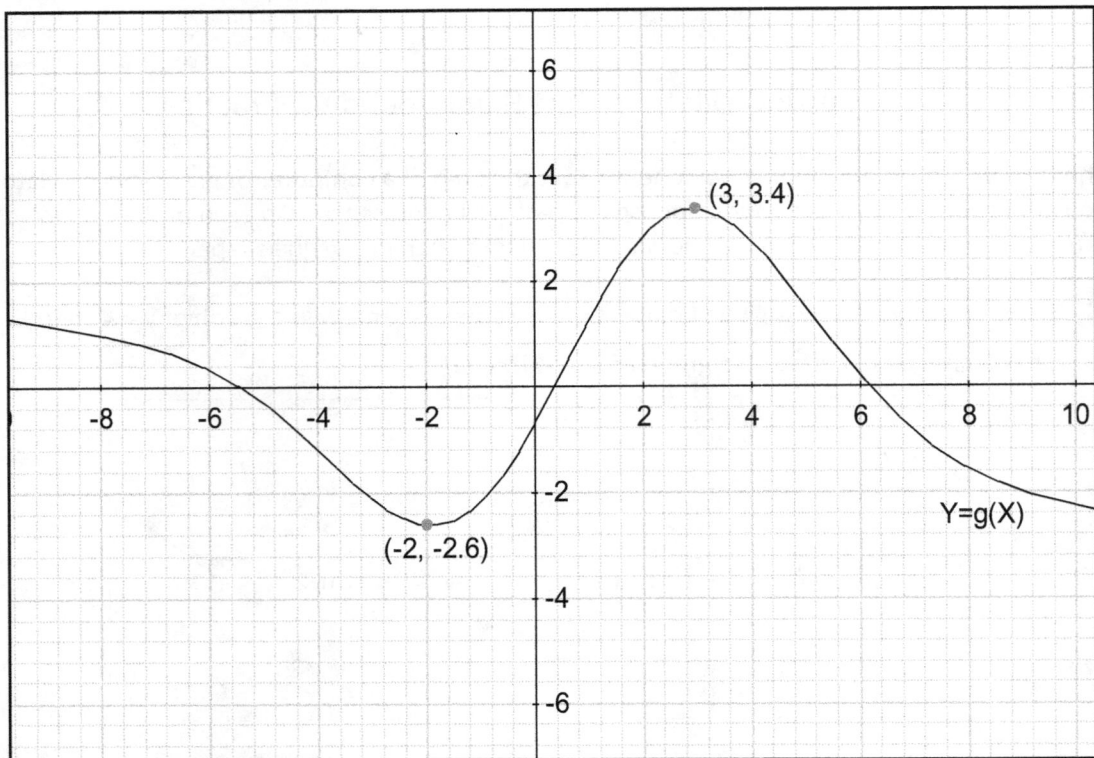

7) Now draw in a graph of the function y = -1*g(x) above. Label and/or color-code your graphs.

8) Graphed below are two functions, h(x) and j(x), which have a simple relationship to each other. One is a vertical dilation of the other. Write out that relationship two ways, by filling in the blanks of the following equations:

$$h(x) = \rule{2cm}{0.4pt} * j(x)$$

$$j(x) = \rule{2cm}{0.4pt} * h(x)$$

y = h(x)　　　　　　　　　　　　y = j(x)

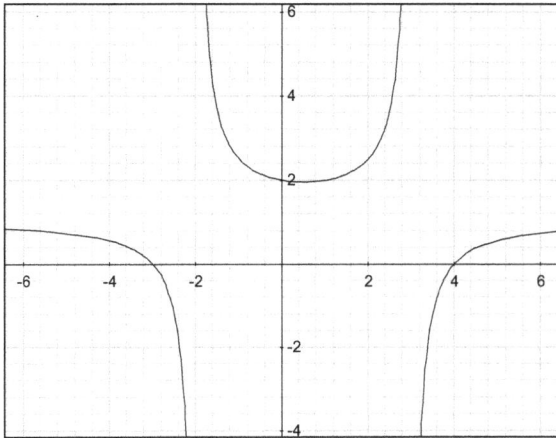 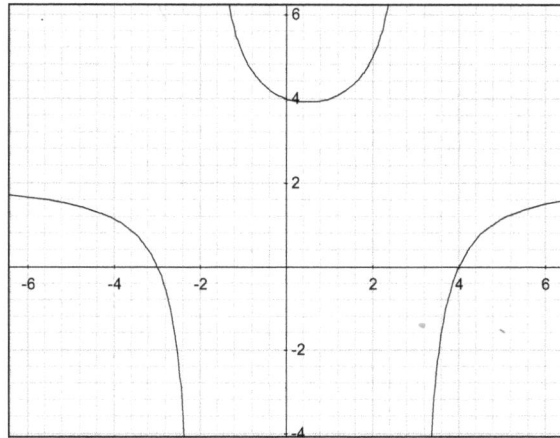

9) Graphed below are two functions, k(x) and m(x), which have a simple relationship to each other. Write out that relationship two ways, by writing two equations.

k(x) =

m(x) =

y = k(x)　　　　　　　　　　y = m(x)

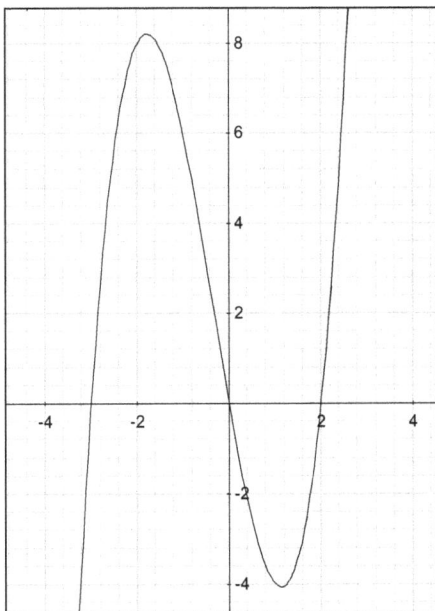 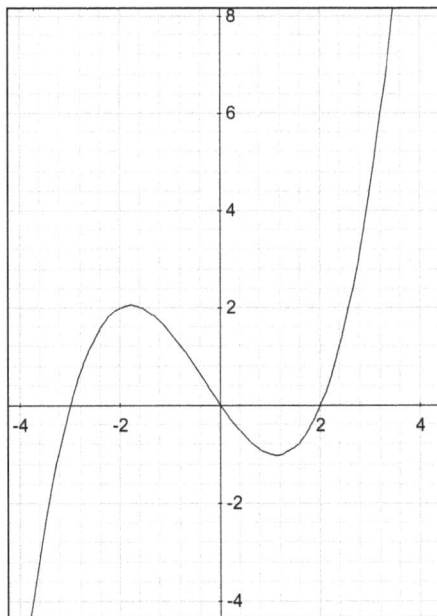

Lesson 3: Combined Vertical Transformations

Learning Objectives

This lesson combines vertical translations and dilations in several quadratic and inverse variation modeling applications.

Math Objectives

- The student will combine vertical dilations and translations in function families $y = ax^2 + b$ and $y = \dfrac{a}{x} + b$.

- The student will use the ideas of translation and dilation to solve applied problems – both graphically and symbolically.

- The student will review geometry and measurement concepts related to areas before and after dilations.

Technology Objectives

- The student will become proficient with the function feature of GX, and with using the variables tool to do dynamic investigations.

- The student will construct dynamic dilations of figures.

Math Prerequisites

- The student must understand vertical translations and vertical dilations.

- Students must know what a function is, and be familiar with two parent functions: $y = x^2$, $y = \dfrac{1}{x}$.

Technology Prerequisites

- Students should have a basic familiarity with Geometry Expressions.

Materials

- A computer with Geometry Expressions for each student, or pair of students.

Overview for the Teacher

The main purpose of this lesson is to combine the transformations of the last two lessons: vertical translations and vertical dilations. Since this is done in context, it was necessary and beneficial to review some concepts related to how linear changes affect area. Specifically, students will either discover, review, or re-discover the principle that if a plane shape is dilated by a factor of k, the area increases by a factor of k^2. While this is somewhat of a digression from the main purposes of the unit, the principle is so commonly mistaken by students that extra review in a relevant context seemed appropriate. Furthermore, it allows for students to make a meaningful connection between geometry, measurement, and function-based algebra in a single scenario.

Part A should be strongly guided, adjusting to the levels of the students. There is a relevant extension prior to problem #7 which should be given consideration. Students should become more independent in problems 7 – 10. Part B builds on the learning from part A, and is sufficiently scaffolded so that students should be able to complete it with minimal input from the teacher.

The beginning of the lesson should flow fairly smoothly, as detailed directions are given to the students. If students use the **line segment** tool instead of the **polygon** tool, they will have to select all four sides and use **construct polygon** before they can calculate areas.

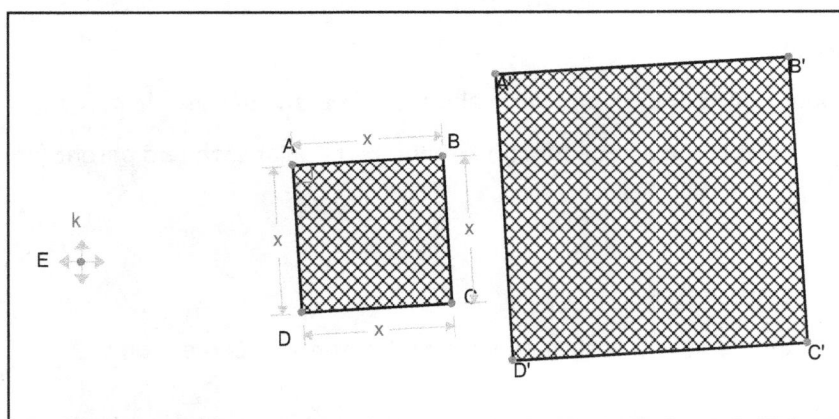

Students may want to assign integer values to x and k to make the patterns easier to see before they make their conjecture. Try to make sure students use the **calculate real area**, not **calculate symbolic area** tool before they answer question #1. The symbolic version may short-circuit some of the thinking and exploration we want them to do here.

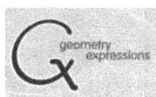

1) Answers will vary – it's a conjecture

2) The area of the second square is k^2 times as great as the area of the first square.

3) See attached page for graphs

4) $y = 4x^2$

5) $y = x^2 - 4$

6) A vertical translation of –4 (down four units)

Verify that students get #4-6 correct as they are working, since these represent key ideas that are fundamental to what comes later.

*** Extension possibility ***

Before students start on problem 7, you may want to give students the following extension to help solidify the concepts.

Have students use GX to construct the models with the statue area indicated. This should be done two different ways – modeling the differences in the main lesson. In the first possibility, they simply construct a 2 – unit square in the corner of both squares as they exist.

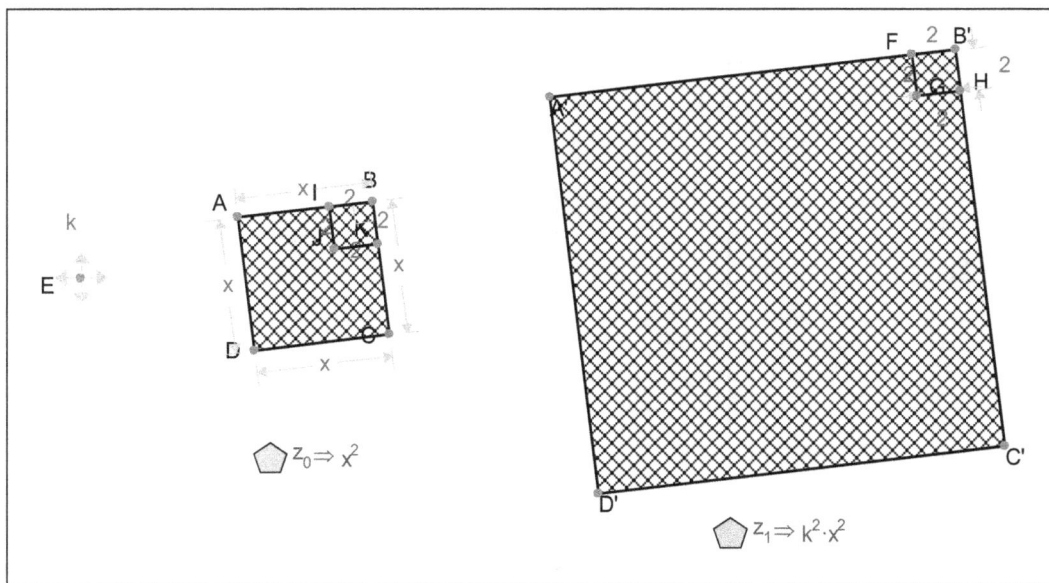

In the second possibility, they will need to delete the second square and the dilation. With the 2-unit square in the corner of the remaining square, they select that whole drawing, and re-create the dilation. This time the statue area gets dilated along with the courtyard. Notice that this sequence on the computer exactly mirrors the order of operations described in the algebra of the main lesson. It also directly mirrors the order of transformations of the function graph.

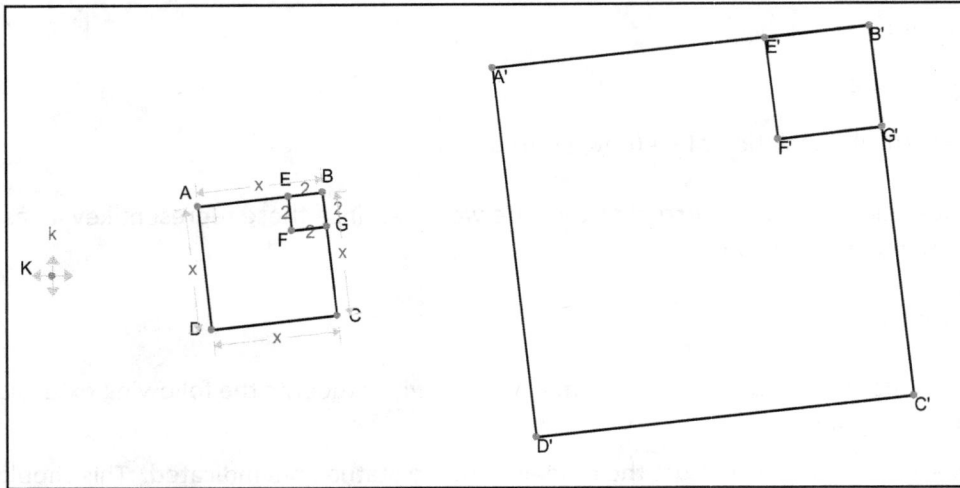

Grids for problem 3 and following:

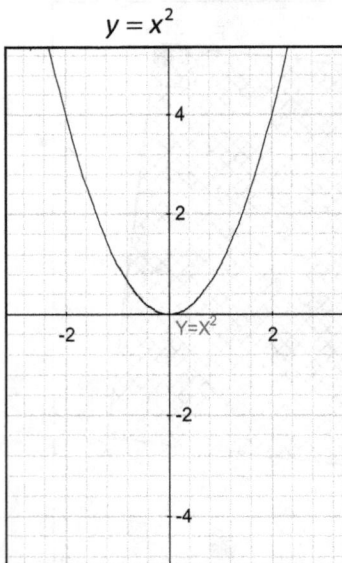

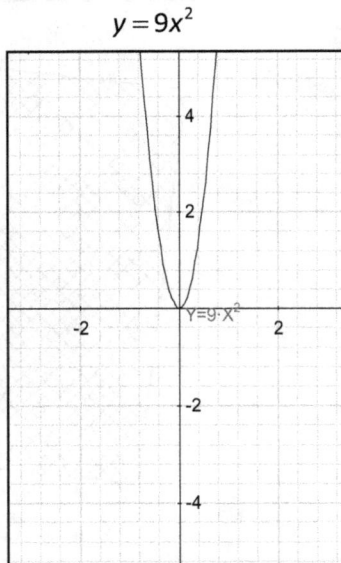

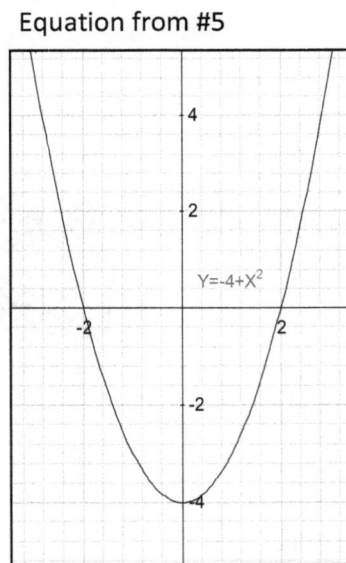

$y = x^2$ $y = 9x^2$ Equation from #5

$y = 9x^2 - 4$

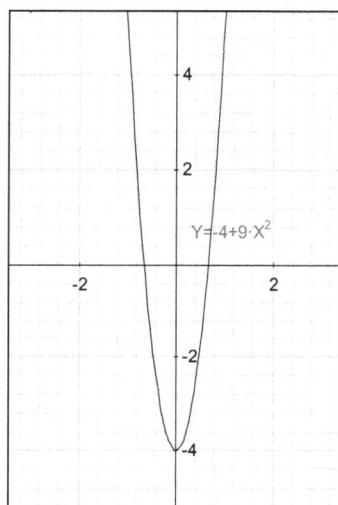

$y = 9(x^2 - 4)$ or $y = 9x^2 - 36$

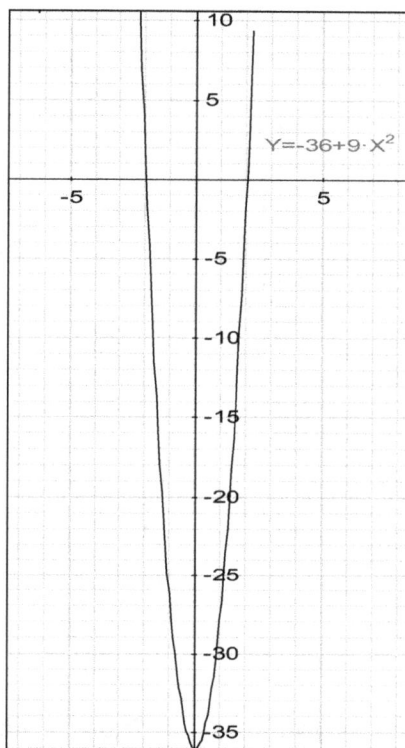

$Y = -4 + 9 \cdot X^2$

$Y = -36 + 9 \cdot X^2$

7)

A) $y = 4x^2 - 25$

B) This graph requires a vertical dilation with a scale factor of 4, followed by a vertical translation of −25 units. Students may put this in more casual terms ("Stretch the graph up and down, four times as tall, then slide it down by 25") which are correct. The teacher should determine how precise the vocabulary needs to be, and communicate this to students.

8)

A) $y = 2.25x^2 - 6$

B) A vertical dilation with scale factor 2.25 followed by a vertical translation of −6

9)

A) $y = 6.25x^2 - 28$

B) A vertical dilation with scale factor of 6.25 followed by a vertical translation of −28.

7)

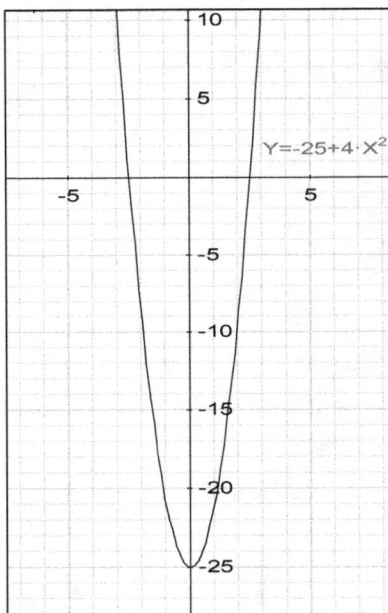

$Y=-25+4\cdot X^2$

8)

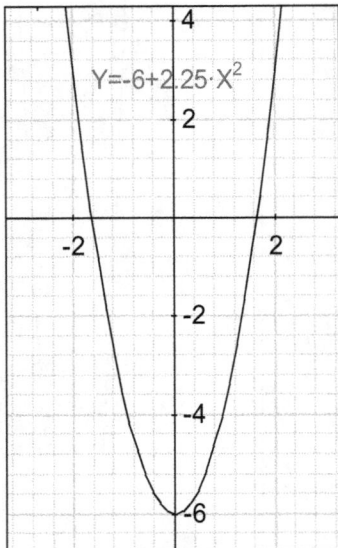

$Y=-6+2.25\cdot X^2$

9)

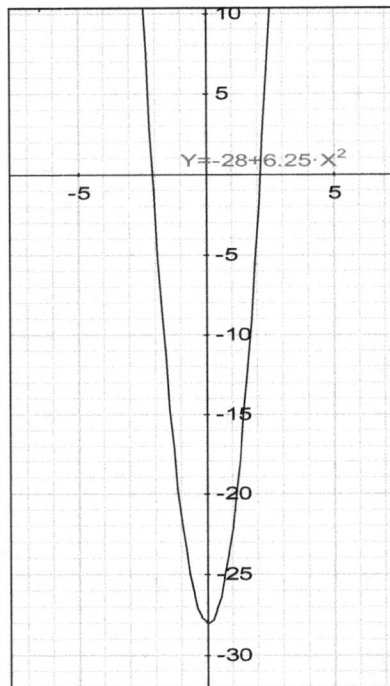

$Y=-28+6.25\cdot X^2$

10)

A) Answers may vary slightly. The most obvious possibility is a courtyard with four courts on one side, and a 7ft. by 7 ft. statue or fountain somewhere inside it.

B) A vertical dilation with scale factor of 16 followed by a vertical translation of −49.

Combined Vertical Transformations/Applications B

This portion of the lesson could be done independently. Students are expected to understand the main ideas at this point, and be able to apply them to a new situation. Also, there is sufficient scaffolding built into the problems to help struggling students adapt. Use interventions as necessary for individual students.

The main difference in the model for problems 1-4 is that the lead coefficient is negative, which indicates a reflection through the x-axis as well as a dilation. Also, students are expected to remember the vertical dilation and translation principles without seeing them visually for the first three problems.

1)   Reflection through the x-axis, vertical dilation of scale factor 4.9, and a vertical translation of 100.

Alternately, a vertical dilation of scale factor –4.9 is technically equivalent to the reflection and dilation mentioned above. However, if a student writes this, he/she is probably simply substituting numbers into the sentence he/she wrote earlier rather than actually thinking about the changes to the graph. The teacher should ask this student to verbally explain what he/she means by a dilation of –4.9. Again, students may want to use casual language to describe the changes ("Flip the graph, stretch it up and down 4.9 times as tall, and then slide it up 100 units.") The teacher needs to decide if precise mathematical vocabulary is going to be required at this point. The casual verbiage is often more meaningful to the student in thinking through the actual changes to the graph. One effective strategy is to have students answer the question twice – once in casual terms, and once using precise mathematical vocabulary. This double-reinforces the concept as well as serving as an embedded vocabulary exercise. Of course, it is difficult to convince students to do this.

2)   A) $y = -4.9x^2 + 300$

        B)   A reflection across the x-axis, a vertical dilation of scale factor 4.9, and a vertical translation of 300.

3)   A) $y = -16x^2 + 450$

        B)   A reflection across the x-axis, a vertical dilation of scale factor 16, and a vertical translation of 450.

4)   A)   -0.4 represents the effect of the force of gravity on planet Cosmotim – derived from a force of gravity of $0.8 \dfrac{m}{\sec^2}$; 30 represents the initial height.

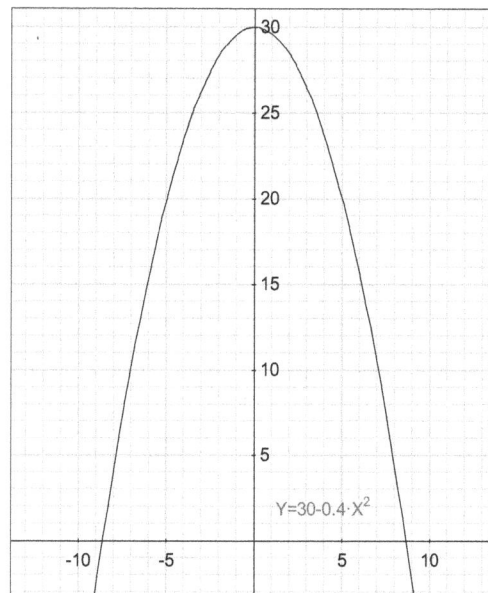

$Y = 30 - 0.4 \cdot X^2$

B)   -0.4 is a vertical dilation. The negative means the graph will be reflected across the x-axis, the magnitude 0.4 is the scale factor of the dilation (i.e. the result will be 0.4 times as tall as the original). 30 represents a vertical translation – shift the graph up 30 units.

5)   A)   Verbally: Answers may vary slightly: A horizontal line through the origin, and a vertical line through the origin. Or the x and y axes.

             Symbolically: $y = 0$; $x = 0$

   B)   A vertical dilation with scale factor of 40.

   C) $y = \dfrac{40}{x} + 1$ or $T = \dfrac{40}{v} + 1$

   D)   A vertical translation of 1 unit.

   E) $x = 0$; $y = 1$;   or (verbal answers may vary slightly) a horizontal line one unit above the x axis, a vertical line through the origin.

6)   A) $y = \dfrac{100}{x} + 1.5$

   B)   6 workers. Students may put down 5.405, but workers only come in whole numbers. Also, if they simply put 5, they've probably ignored the instructions in problem 5.

   C)   Verbally: a horizontal line 1.5 units above the x-axis, a vertical line through the origin.

             Symbolically: $x = 0$; $y = 1.5$;

   D)   A vertical dilation with scale factor 100; a vertical translation of 1.5.

5A)

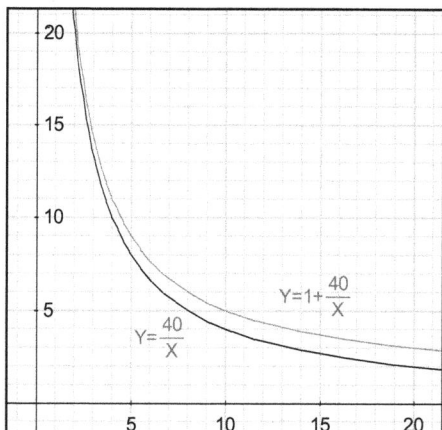

5B)

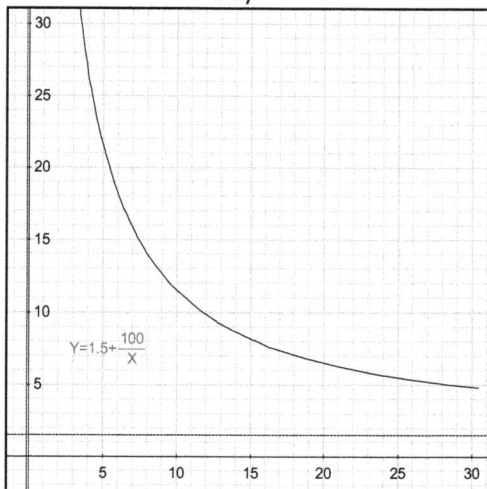

Student Worksheets

Student worksheets follow

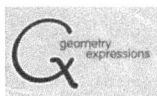

Name: _____

Date: _____

Combined Vertical Transformations/Applications A

In the previous lessons, you studied how adding a constant to a function caused a vertical translation (slide), and how multiplying a function by a constant caused a vertical dilation (stretch). Today, you are going to combine the two processes and solve some basic application problems.

The first set of activities relates to the problem of designing a student courtyard which will have one or more basketball courts on one side and a statue or fountain inside it. The question will relate to how the length and number of basketball courts will affect the area students have in the courtyard to socialize. To begin with, we'll review some geometry and measurement concepts you are probably already familiar with.

Your first job is to open a new GX file, and construct a square with side length X. Toggle the axes and grid lines off (top tool bar icon:). Use **draw polygon** to make a quadrilateral. **Constrain** each side **length** to be x , producing rhombus. Since opposite angles of a rhombus are congruent, you only need to constrain one angle to perpendicular to produce a square. Select two adjacent line segments at the same time by holding the shift key, and use **constrain perpendicular** .

Now you are going to create a scale model of your square by doing a geometric dilation. This will stretch in all directions, so length and height of the object is affected.

Click and drag a box around your drawing to select everything, then use **construct dilation** . Click to place the reference point on your page somewhere outside your square, and name the scale factor k. A second square, which is a scale model you your first one, should appear.

Take a couple minutes to click and drag on the various elements of your drawing to get a feel of how they relate to each other. Make sure you investigate the effects of changing both k and x. Go to the Variables tool panel, highlight the variable, and adjust its value either by typing in numbers or dragging the scroll bar. You may need to scale in or out (top tool bar icons , and -- like the zoom function on Word) at various times to see the diagram well.

Now we're going to compare the areas of the two squares. Highlight the space inside each square, and use **calculate real area** . There is a predictable relationship between these areas based on the value of the scale factor k. If you think you remember it, test your idea with various number values for k. If not, then test various number values for k and formulate a conjecture.

1) Your Conjecture in words:

_____

_____

Another way to test your conjecture is to use the **calculate symbolic area** tool [x']. Note that you will need to delete your other area measurements first.

2) The area relationship given symbolically:

_____

Why does this work? The relationship between changing lengths and changing areas works for any flat shape. Think of a rectangle with length a, and width b. Its area is $a*b$. If we do a geometric dilation with a scale factor of 3, we stretch both the length and the width. We now have a rectangle with length $3a$ and width $3b$, which yields an area of $3a*3b = 9ab$. This is 3^2 times the area of the original rectangle.

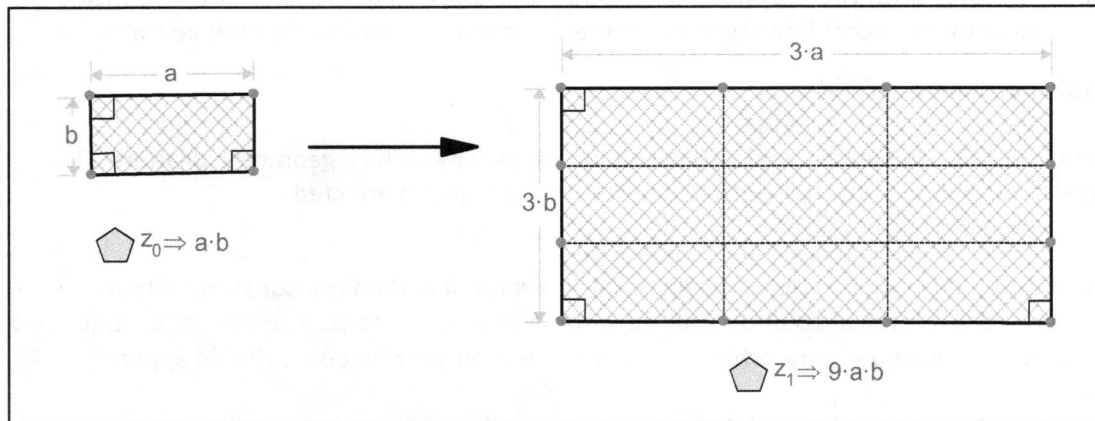

3) Now we are going to consider the case where $k = 3$, and examine the problem algebraically. It should be fairly obvious that the area of both squares depends on the length of x. This allows us to write two function equations: $A = x^2$ for the first square, and $A = 9x^2$ for the second. Open a new file in GX, toggle the axes and grid lines back on, and use the function tool to graph $y = x^2$. Note that the y values in the graph are the areas corresponding to any given length of x in our construction. Negative values for x have no meaning in this context. A vertical dilation, then, corresponds to multiplying the area of a square by a constant. In our case, the area of the second square is always 9 times the area of the first. Graph the function $y = 9x^2$ and see if the graph behaves as you would expect it to.

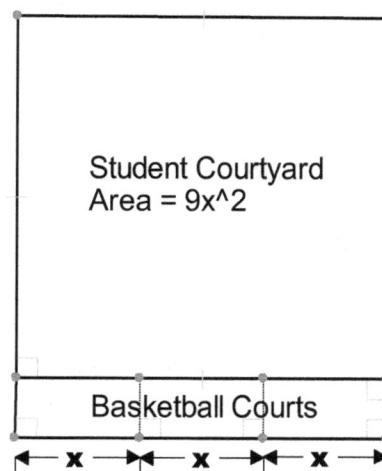

Copy your graphs onto the axes on the last page as you go.

Application:

Suppose you are going to build a square student courtyard, with a basketball court along one side. The length of a basketball court is x, and the area of the courtyard is given by $y = x^2$ in our first example. Similarly, the $y = 9x^2$ model would relate to the larger option of the courtyard in which three courts were on one side.

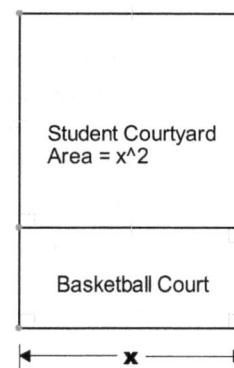

4) What equation would express the area of a courtyard with two basketball courts on one side?

There will be a statue with a square base in one corner of the courtyard, which then must be subtracted from the total area. In our first example, the statue would be 2 ft. by 2ft., for a total of 4 ft^2 lost. (Still one basketball court length.)

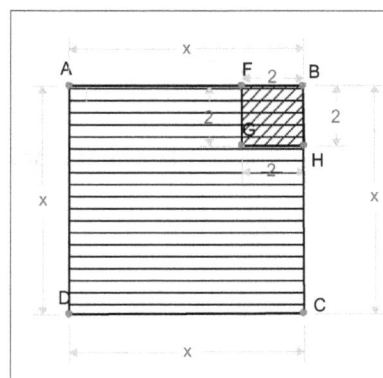

5) What equation models the courtyard area for this scenario?

6) What change in the graph of the function corresponds to the change you made to the equation for #5?

Graph your new equation in GX to check your idea. Copy it down on the last page.

When it comes to the scaled-up version of the courtyard, we have two options.

A) We could scale up our courtyard, then take out the area for the 4 ft^2 statue. This would result in the equation $y = 9x^2 - 4$ and correspond to a vertical dilation by a factor of 9, followed by a vertical translation of –4.

B) We could put the statue into our plans, then scale everything (including the statue) up proportionally. Now the statue base would be 6 ft. by 6 ft., or 36 ft^2. This would result in the equation $y = 9(x^2 - 4)$, and correspond to a vertical translation of –4, followed by a vertical dilation by a factor of 9. Note that this is equivalent, by the distributive property, to an equation of $y = 9x^2 - 36$.

Graph both functions in GX, and copy onto the grids on the last page.

You are going to look at several variations on this courtyard problem. For the first three, give the equation, sketch the graph, and describe the transformations (vertical translations and dilations) required to produce the graph from the parent function of $y = x^2$. In each case, the courtyard is square and x will indicate the length of one basketball court. Also, the dimensions of the statue or fountain given are final dimensions, after any scaling is done.

7) The courtyard has 2 basketball courts on one side, and a 5 ft by 5 ft fountain in the middle.

   A) Equation:

   B) Describe the transformations:

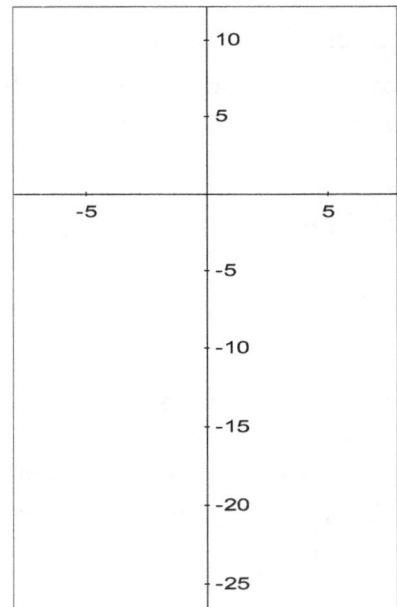

8) The courtyard has 1.5 basketball courts on one side, and a statue with a base that is 2ft. by 3 ft.

  A)   Equation:

  B)   Describe the transformations:

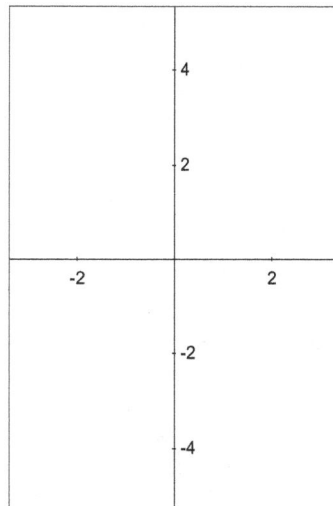

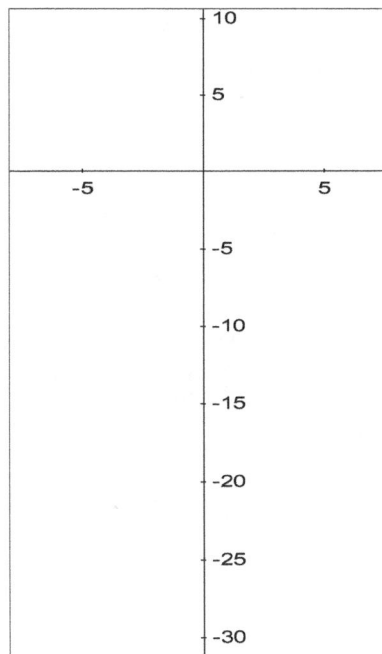

9) The courtyard has 2.5 basketball courts on one side, and a fountain that measures 7ft. by 4 ft.

  A)   Equation:

  B)   Describe the transformations:

10) For the same courtyard problem, an equation is
    $y = 16x^2 - 49$.

  A)   Describe one situation that could be represented by that equation.

  B)   Describe the transformations:

Grids for problem 3 and following:

$y = x^2$

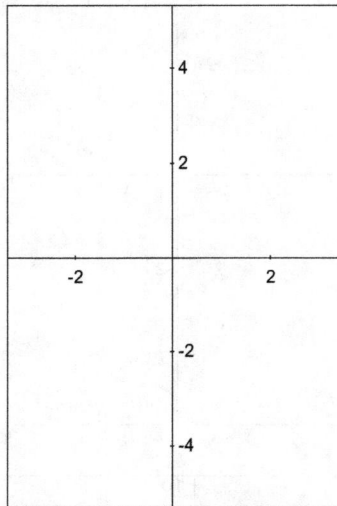

$y = 9x^2$

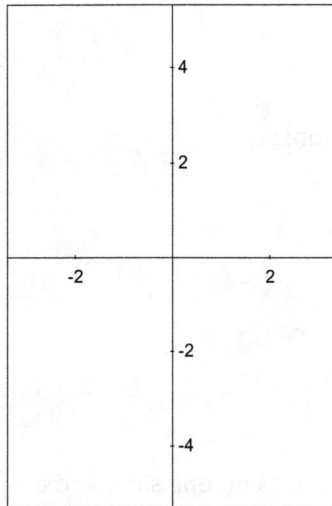

Equation from #5:

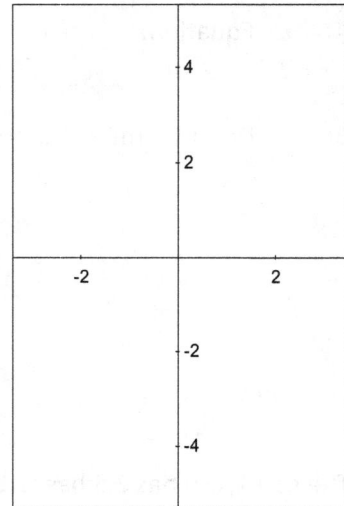

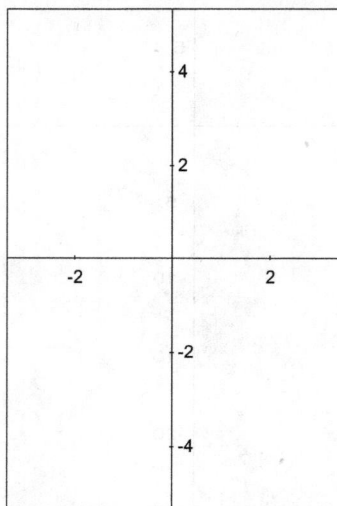

$y = 9x^2 - 4$

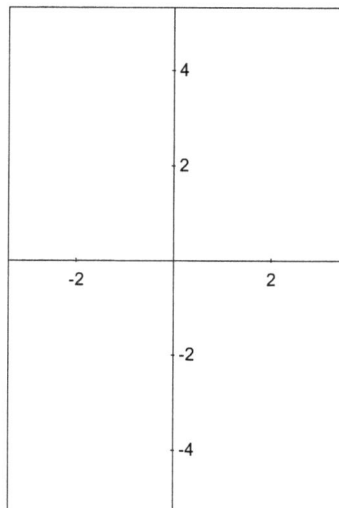

$y = 9(x^2 - 4)$ or $y = 9x^2 - 36$

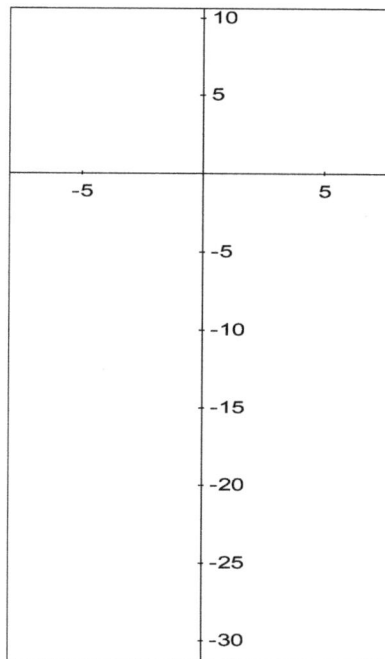

Name: _____

Date: _____

Combined Vertical Transformations/Applications B

If you graph the height of a falling object (ignoring air resistance), the result is the shape of a parabola, which can be modeled with a function in the family whose parent is $y = x^2$. Time is the independent variable (x) and height is the dependent variable (y).

The height of an object falling from 100 meters up would be given by the equation $h = -4.9t^2 + 100$, where h is the height, measured in meters, and t is the time, measured in seconds. The effect of gravity is given by the -4.9; this is derived from the force of gravity, which is $9.8 \dfrac{meters}{second^2}$.

1) Describe the vertical transformations necessary to produce the graph of the example function, starting with the parent function of $y = x^2$.

2) A second object is dropped from a height of 300 meters.

   A)   Write an equation for the height of this object as a function of time.

   B)   Describe the vertical transformations necessary to produce the graph of this function, starting with the parent function of $y = x^2$.

3) If we measure in feet, the effect of gravity is given by the coefficient -16 (derived from 32 $\dfrac{feet}{second^2}$). An object is dropped from a height of 450 feet.

   A)   Write an equation for the height of this object as a function of time.

   B)   Describe the vertical transformations necessary to produce the graph of this function, starting with the parent function of $y = x^2$.

4) Planet Cosmotim is smaller, and has less gravity than earth. The equation for a particular falling object on that planet is $h = -0.4t^2 + 30$.

A) Sketch a graph of this situation.

B) Explain what the numbers −0.4 and 30 mean in this context.

C) Explain what the numbers −0.4 and 30 mean in terms of transformations of the parent graph $y = x^2$.

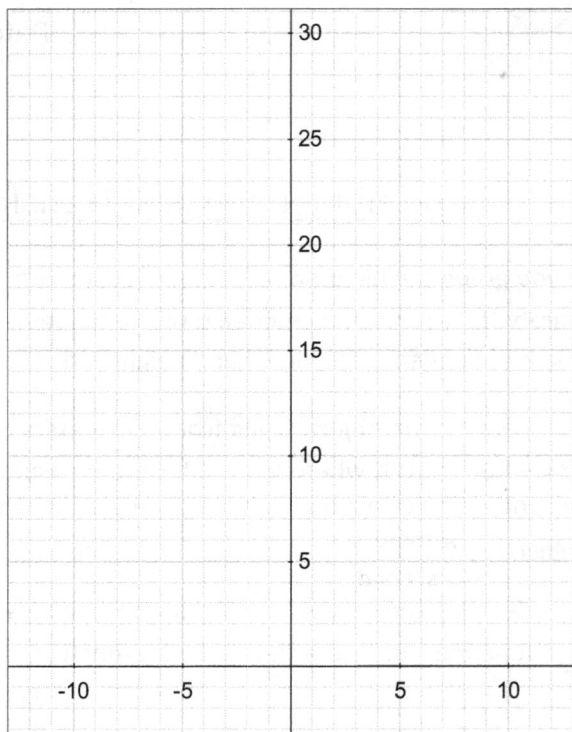

Inverse variations are modeled with equations from the family of $y = \dfrac{1}{x}$. In general, as the independent variable (x) increases, the dependent variable (y) decreases at a decreasing rate. These models are common when there is a set amount of something being divided equally into a variable number of parts.

5) Your school ASB is organizing a campus clean-up day as part of their community service efforts. They estimate that it will take around 40 hours of labor to get the school as clean as they hope. When they reach their goal, there will be a short celebration. Obviously, more volunteers would mean less time to wait for the party. The work is evenly divided among the volunteers, which gives the inverse variation equation $T = \dfrac{40}{v}$. Since it is impossible to have a negative number of volunteers or a negative amount of time used, we will only look at the first quadrant for this and similar problems. Also, the mathematical model is continuous, allowing values for v like 1.27. However, since it is impossible to have 1.27 volunteers in real life, you must interpret your results in ways that make sense.

A)   Draw a graph of the function $T = \dfrac{40}{v}$.

B)   Describe the asymptotes in your graph – either verbally or symbolically.

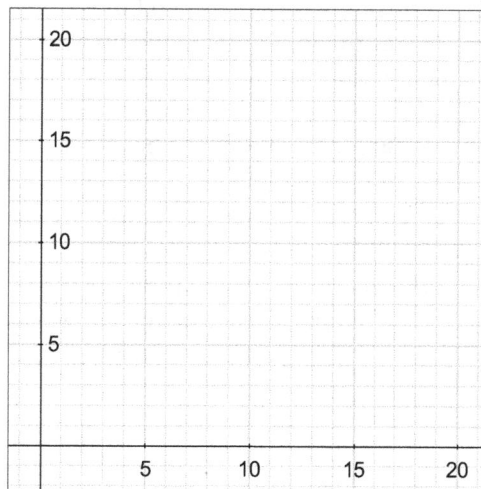

C)   Describe the transformation necessary to get the graph you drew from the graph of the parent function $y = \dfrac{1}{x}$.

D)   The principal decides that she needs to meet with the volunteers for 1 hour before they get started cleaning. This, of course, adds to the total time it takes to finish the job – every scenario will be 1 hour longer. Write a new equation for the situation, and graph it on the same axes above.

E)   What additional transformation did your change in part D cause?

F)   Where are your asymptotes now?

6) A business is planning to build a new building on what is currently a vacant lot. They need to get the job done in one weekend, and are hiring local teens to do the clean-up work. They don't know how many of the teens they've interviewed will show up, but once they are there, they will work until the job is done in order to get paid. They estimate that there is 100 hours of work to be done. In addition, everyone is required to attend a 1.5 hour meeting to go over safety and procedures. The amount of time it takes to get finished obviously depends on how many workers they have.

A) Write an equation for the time it takes to complete the work as a function of the number of workers.

B) Graph the function to the right.

C) Assuming no one will work more than 20 hours on the weekend, what is the minimum number of workers the company needs in order to get done?

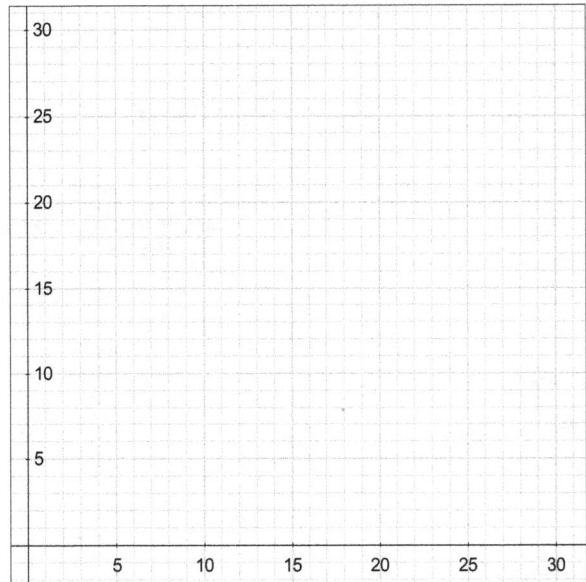

D) Describe the asymptotes in your graph – both verbally and symbolically.

E) Describe the transformations necessary to get the graph you drew from the graph of the parent function $y = \dfrac{1}{x}$.

Lesson 4: Circular and Harmonic Motion

Learning Objectives

The students will apply combined vertical translations and dilations in the context of applications of the sine function. Students will be exploring circular and harmonic motion graphically, symbolically, contextually, and through computer simulation. The examples are simplified in the fact that time is essentially ignored, but height and range of motion are explored in-depth. Adjustments in period length and starting point will be addressed later in the unit, so the examples can be revisited in that more authentic context. We believe students will be more successful at that point after having mastered the vertical components of a sinusoidal curve first.

Math Objectives

- Understand vertical translations and dilations as applied to the sine function. Specifically, students will create equations, graphs, and simulations of sinusoidal curves, given a description of a relevant situation.

Technology Objectives

- The student will create simulations of vertical components of circular and harmonic motion on GX.

Math Prerequisites

- The student must understand the sine curve, vertical translations, and vertical dilations.

- The student must understand radian measure for angles of rotation.

Technology Prerequisites

- Basic understanding of Geometry Expressions developed through earlier lessons in this unit.

Materials

- A computer with Geometry Expressions for each student or pair of students.

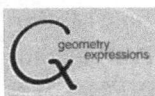

Overview for the Teacher

The lesson starts with a basic sine curve in the abstract, tying circular motion dynamically to the function graph, which is created as a locus in GX. Students then do vertical translations and dilations to it, and apply those ideas to some basic circular motion contexts. Building on that, students modify their motion simulations to create harmonic motion – first connected to a circular pattern, then to a function graph of a sine curve. There is an optional activity which asks students to first produce simulations, then produce equations and graphs based on simulations of other students. In short, the lesson reinforces and practices vertical dilations and translations as applied to sine curves, and further relates them to simplified realistic contexts and meanings.

1) This first exercise reviews the sine curve as a record of vertical motion of a point around the unit circle. It also introduces the idea of a locus and sets students up for the more advanced modeling and simulations they will be doing.

When students construct the locus, GX will choose values for t which match those in the variables window by default. These can be changed in the locus pop-up.

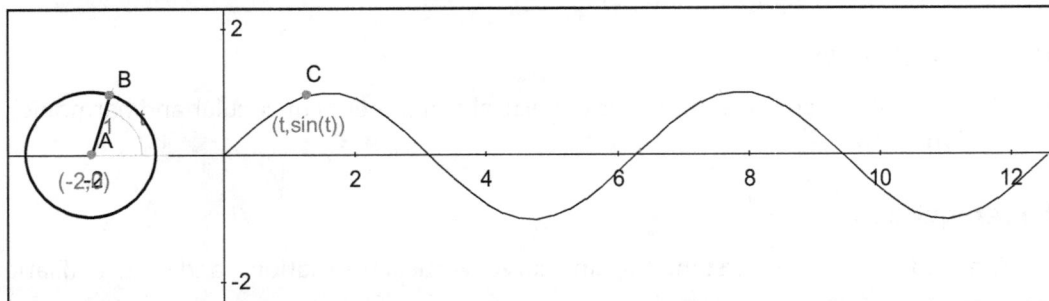

You may want to discuss the location of key points with students in terms of fractions of π, since those intervals aren't readily obvious to some students, and the axis scale is in integers. Many of them will, at least mentally, need to translate this to degrees to make good sense of the pattern.

Of course, you can add more teacher checkpoints to the lesson, but this one is fairly important to make sure students are on the right track.

2)

A) $y = 2\sin(t) + 3$

B)   Remind students that they must type in the * to indicate multiplication in GX. Students should be able to produce the modified model with only a couple of changes to values. Encourage them <u>not</u> to start over from scratch.

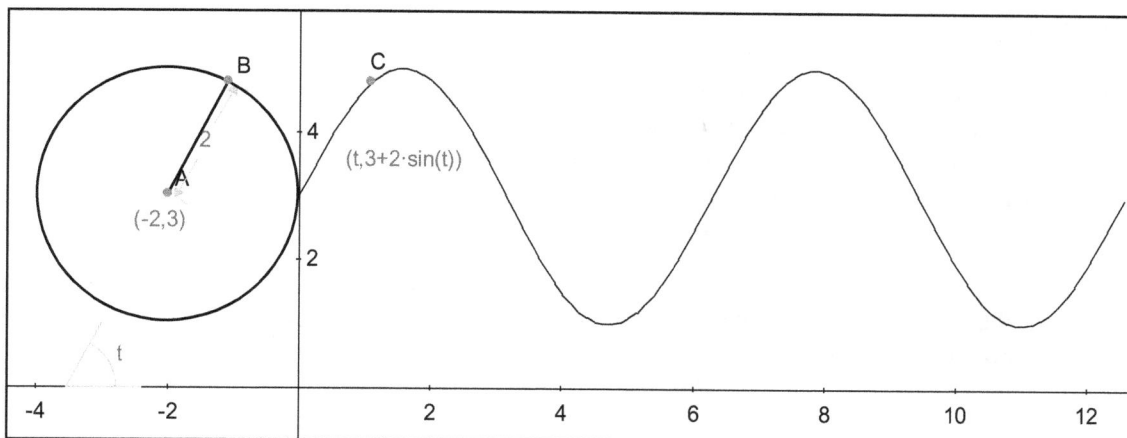

$(t,3+2\cdot\sin(t))$

C) A vertical dilation of the function corresponds with a proportional stretch of the radius of the circle.

D) A vertical shift of the function corresponds with a vertical shift of the circle.

3)

A) A vertical dilation of magnitude 5, and a vertical translation of 6 units (up).

B) $y = 5 \sin(x) + 6$

C)

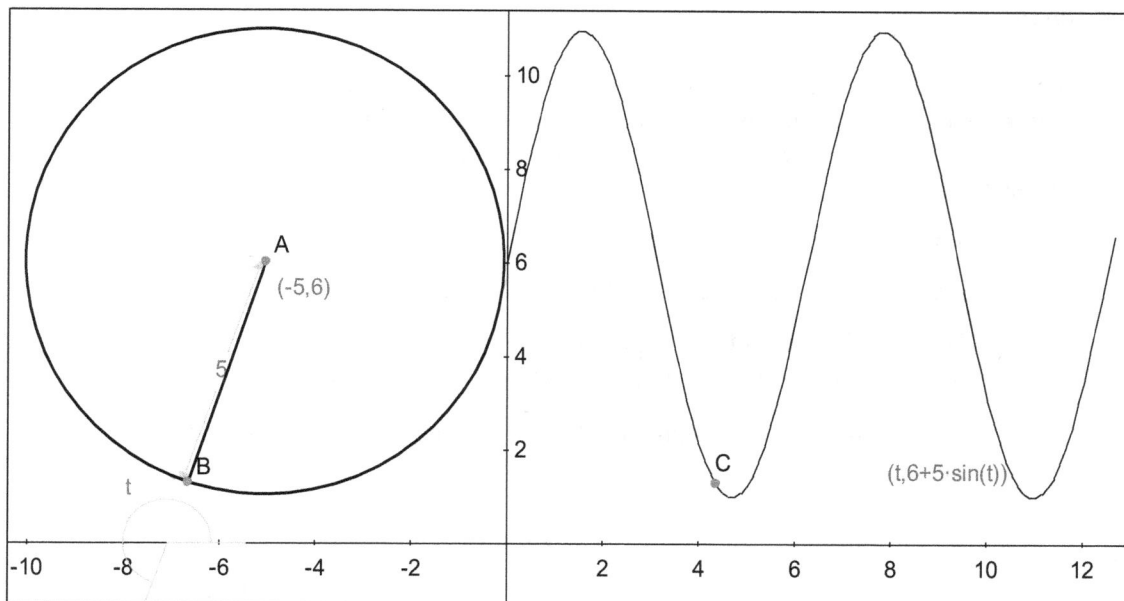

$(t,6+5\cdot\sin(t))$

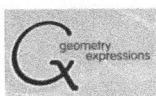

Harmonic Motion:

Students can quickly see how harmonic and circular motion correspond with each other. We highly recommend doing the following more detailed demonstration model for the class. Some students will find this more visually intuitive for the pattern described.

- Plot a point directly above your point C, constrain its position, and connect it to C with a line segment.

- Make this segment a dashed line to represent the spring by highlighting it, right clicking, and using properties/line style.

- Draw a horizontal line segment that goes through that point to represent the board.

- Draw a small circle with center on point C, and constrain its radius.

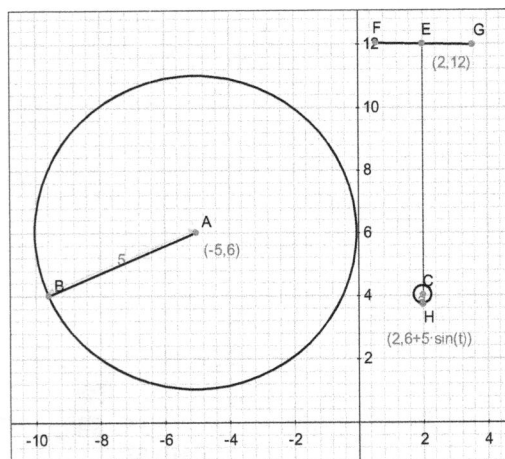

While this adds nothing mathematically to the drawing, when you animate it, it looks like a weight suspended from a spring beneath a board. Students may want to duplicate this; exploring how to do so independently may help them to become more familiar with the software if they have extra time.

4) One thing to watch for is now we are translating the sine curve down, instead of up. Also, the total range of motion is 12 inches or units, which makes a dilation of magnitude 6. Students will need to type in a range of values for t: either 0 through 6.28 or 0 through 12.56.

A) $Y = 6*\sin(t)-10$

B)   A vertical translation of –10 (down ten units) and a vertical dilation of scale factor 6.

C)   The coordinates of the animated point should be some negative constant for x (or x_0 with a negative value in the variables window), and $6*\sin(t)-10$ for y.

D)   The whole GX screen is given below. Students will only have the function copied, not the points or coordinates.

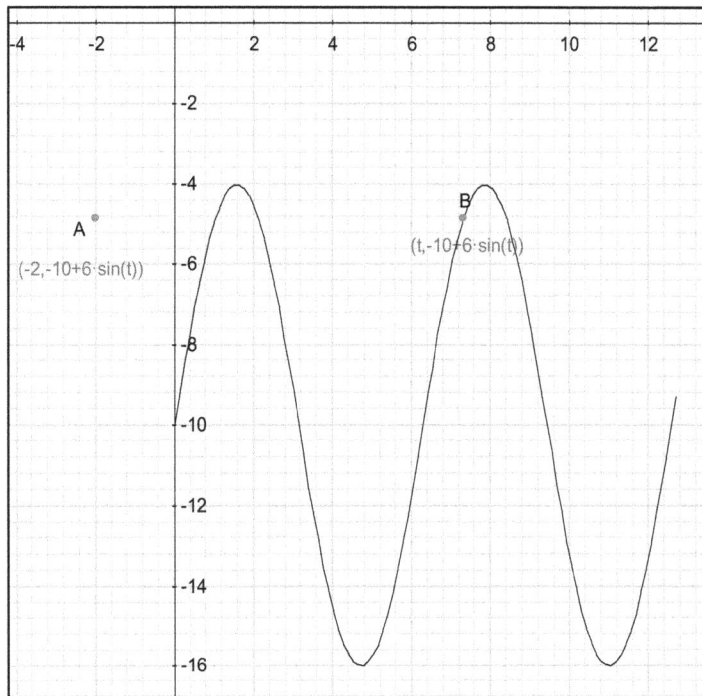

5)

A) $y = 3.5\sin(t) - 8$

B)   A vertical dilation w/ a scale factor 3.5, and a vertical shift of −8 (8 units down).

C)   The coordinates of the animated point should be some negative constant for x (or x_0 with a negative value in the variables window), and 3.5*sin(t)-8 for y.

D)   The whole GX screen is given below. Students will only have the function copied, not the points or coordinates

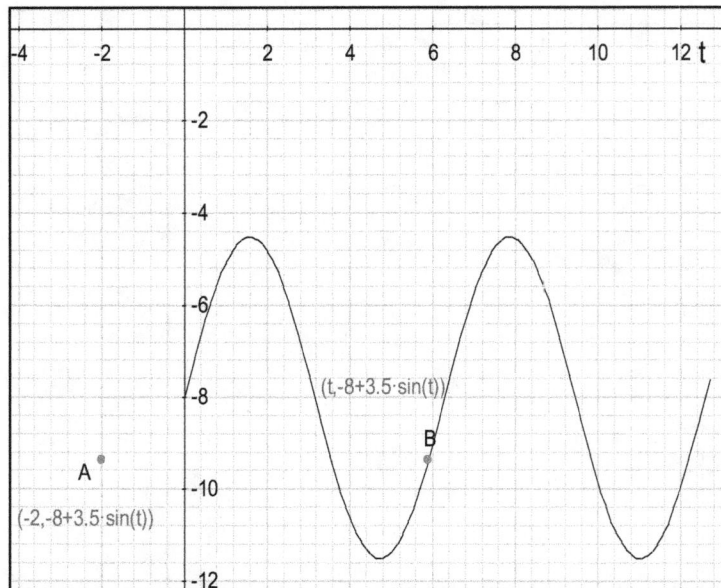

Classmate Simulation Challenge:

Obviously, answers will vary. This activity is optional, but may be a motivating way for students to get some practice done. It can be expanded to have students attempt more than two of their classmates' models, or possibly made into a game or contest. This all depends on time, computer access, student interest, and the instructor's classroom management style. A similar activity will follow the later lesson and incorporate phase shifts and period changes.

Practice:

Students should do independent practice writing equations, graphs, and verbal situations. The "Classmate Challenge" is a good way to get this practice started. Since such exercises are fairly standard and easy to obtain, we don't include a full set here. Some ideas for teacher-created exercises are given below. In all cases, the period must remain 6.28, and the object in question must start out at the starting point of the natural sine curve. These constraints limit the range of the examples and force them to be somewhat contrived, but that will be largely remedied in the later lesson.

- A Ferris wheel has a radius of 45 feet and is centered 55 feet above the ground. Find the height of Tom after t seconds. (Modify the numbers as many times as you like, just make sure your height is always greater than your radius.)

- Use the gear wheel of a machine, and set the problem up like the Ferris wheel (but with smaller values and/or different units). This can be set up as a large, industrial machine,

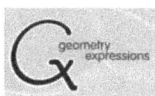

like in the lesson, or a small, precision machine, which might better allow for including decimal values for the dilations and translation.

- Harmonic Motion: Weight on a spring hanging from a fixed point. This one is a little trickier, since the time has to start when the weight is at its point of equilibrium, and on its way up: A weight hangs on the end of a spring below a fixed board. At rest, it is 7 inches below the board. Someone stretches it down to 12 inches below, and lets go. Assume the spring preserves all the energy from gravity, so the motion continues up and down indefinitely. Let time start when the weight is 7 inches below the board and on its way up. Give the equation and the graph of the weight's height as a function of time. Again, modify the numbers as many times as you want.

- Give an equation, and have students write a semi-plausible situation that matches it. E.g. $y = 10\sin(t) + 13 \rightarrow$ "A Ferris wheel with radius 10 meters is centered 13 meters above the ground. The equation gives the height of a person after t seconds."

- Give a graph, and have students produce the equation and/or a verbal situation. These can be created easily in GX with the function tool. Then use Edit/Copy Drawing, open a Word document, and Edit/Paste Special.

  Paste it as an enhanced metafile, and you won't lose any special characters. You can crop, resize, and format the diagram in Word.

Student Worksheets

Student worksheets follow.

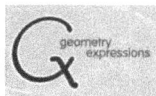

Name: _____

Date: _____

Circular and Harmonic Motion

As you may know, the sine function and its corresponding graphic curve are used to model almost all types of circular motion. They can also be used to model a similar pattern called harmonic motion. In this lesson, you will investigate those patterns, and learn to represent them with equations, function graphs, verbal descriptions, and dynamic simulations using computer software. For now, we will use a simplified version of the patterns in which the cycle repeats itself every approximately 6.28 (2π) seconds or minutes, and the motion always starts at the natural starting point of the sine function. Later in the unit, we'll learn how to deal with changes in these two components, and return to re-investigate our models.

1) Model the basic sine curve:

As you recall, $\sin(\theta)$ corresponds to the y-coordinate of a point on the unit circle, rotated θ radians counterclockwise from the positive x-axis. Open a GX file, and **draw a circle** anywhere with point A at its center and B on the circle; **constrain** its **radius** to be one. For these models, we are not going to be concerned with the x-coordinate on the unit circle, or $\cos(\theta)$. Because of this, we can move the circle to the left of the origin. Constrain the center to a point left of the origin, such as (-2, 0). Draw in line segment AB, and constrain the angle it makes with the x-axis to be t. The angle of rotation is t, which in our simplified models also represents the time point B has been in motion (due to the cycle being 6.28 seconds, as mentioned above.)

Animate the drawing; show two revolutions of point B. To do this, select t in the Variables tool panel, and type in 0 to 12.56 in the boxes at the bottom, then click on the play button.

Now you are going to let the x-axis represent time, and create a function of the height of point B at various times. Draw point C anywhere, then constrain its coordinates to (t, $\sin(t)$). Reanimate the drawing. To see all the possible point C's at once, we create a locus. A locus is a set of all points that meet specific criteria (like having the coordinates we stated.) Click on point C and then **construct locus**. Reanimate the point. What have you created?

_____

Have your teacher check your drawing and animation, and initial here:

_____

2) Modify the sine curve. At this point, you know how to indicate a vertical dilation and translation in a function equation.

   A) Give the equation for y = sin(x) after a vertical dilation of 2 and a vertical translation

        of 3: _____

B) What change to the circle corresponds with a vertical stretch of the function?

C) What change to the circle corresponds with a vertical shift of the function?

D) Now modify your circle and the coordinates of point C to represent the changes you indicated in part A. The animation of your new model should still work. Make sure the circle, point C, the locus (function graph), and the animation all still match each other. Sketch your result below.

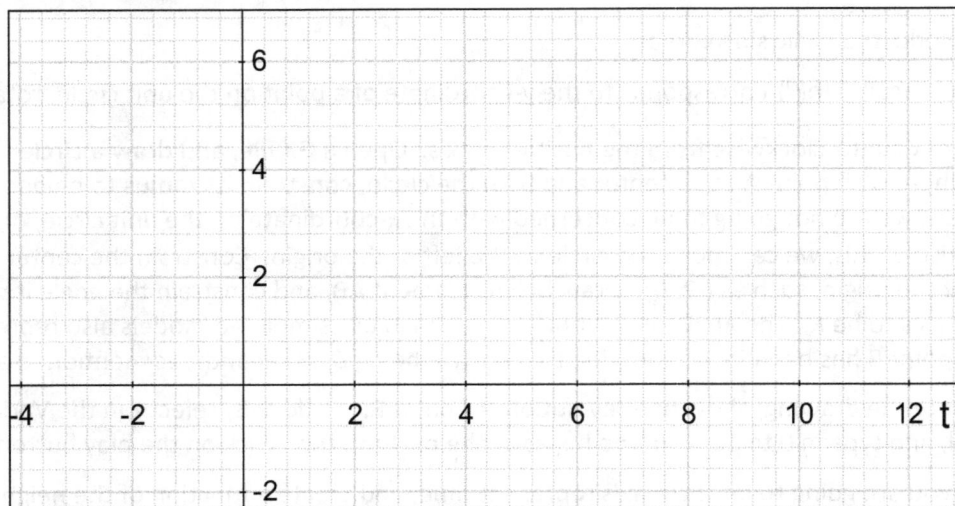

3) Now consider a gear wheel that is part of an industrial machine. It has a radius of 5 feet, and is mounted 6 feet above the floor. (Again, assume a period of 6.28 seconds, and a starting point directly right of center.) We want to know the height of a particular point on the wheel as a function of time.

A) What transformations of the parent function y = sin(x) must be made to give a function equation for this situation?

B) Write the function equation for the given situation.

C) Modify your model in GX to match the situation. You may need to move your circle farther to the left. Sketch your result below.

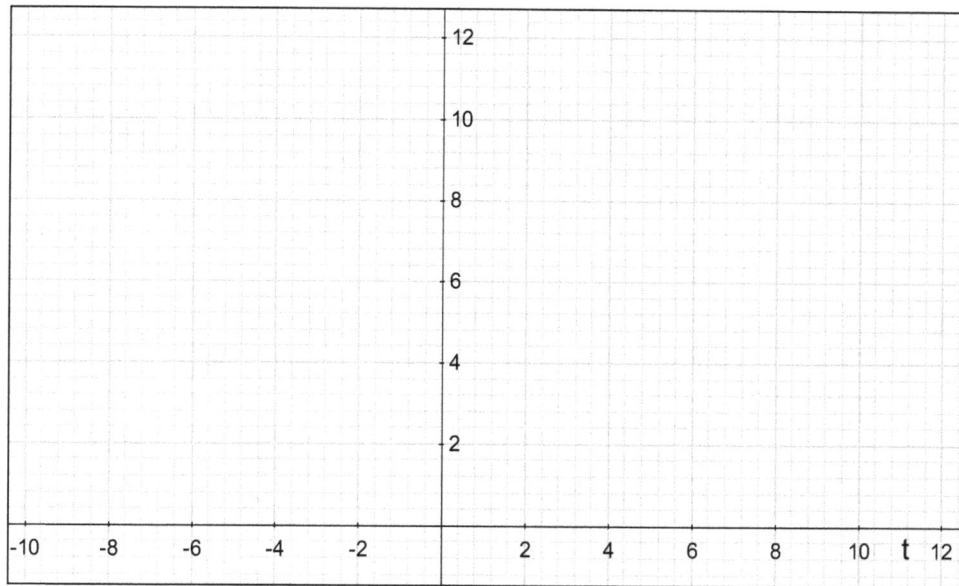

Harmonic Motion:

A closely related pattern is called <u>harmonic motion.</u> To quickly get a picture of this, delete your locus from problem 3, and change the *x*-coordinate constraint on point C to a constant (like 2). Run the animation again. You can create a continuously repeating pattern by changing the animation mode next to the control buttons in the variables tool panel to a double arrow.

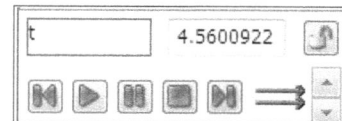

This simulates the action of a weight hanging from a spring below a board. The resting position of the weight – its point of equilibrium – is at 6 inches above the floor. The lowest point it reaches is 1 inch above the floor, and its maximum height is 11 inches above the floor. The spring and the force of gravity counteract each other in a continuous back-and-forth pattern. This pattern would continue indefinitely assuming the spring can capture and release 100% of the energy from gravity.

4) A weight is suspended beneath a table by a spring. Its resting position – point of equilibrium – is 10 inches below the table. It is stretched to 16 inches below, and springs up to 4 inches below, then repeats its motion in a harmonic pattern. Assume it completes a full cycle in 6.28 seconds, and time starts when it is 10 inches below the table, on its way up.

A) Write the equation for this function, with the tabletop as a reference point/height of zero: _____

B)   What transformations must be done to the parent function $y = \sin(t)$ in order to produce this new function?

C)   Letting the *x*-axis represent the table, create a simulation of the weight's motion in a new GX file and animate it. What did you input as the coordinate of the point to accomplish this?

D)   Make a function graph by plotting a point with the coordinates (*t*, your function equation), and constructing its locus. Copy your function graph.

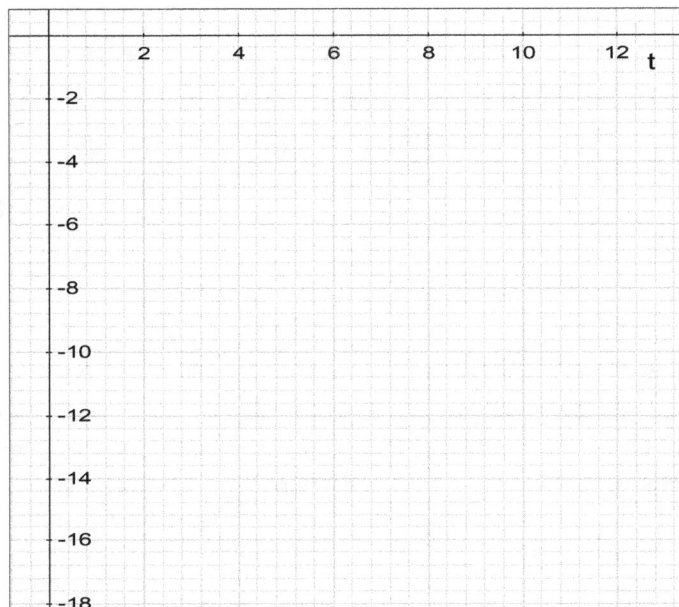

5) A weight is suspended beneath a table by a spring. Its resting position – point of equilibrium – is 8 inches below the table. It is stretched to 11.5 inches below, and springs up to 4.5 inches below, then repeats its motion in a harmonic pattern. Assume it completes a full cycle in 6.28 seconds, and time starts when it is 8 inches below the table, on its way up.

A)   Write the equation for this function: _____

B)   What transformations must be done to the parent function $y = \sin(t)$ in order to produce this new function?

C)   Letting the *x*-axis represent the table, create a simulation of the weight's motion in a new GX file. Place it somewhere to the left of the *y*-axis, and animate it. What did you input as the coordinate of the point to accomplish this?

D)   Make a function graph and copy it here:

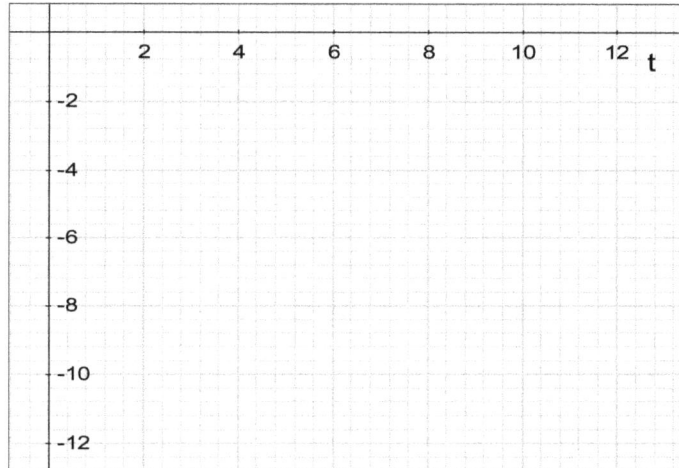

Classmate Simulation Challenge

You are going to create a simulation of either circular or harmonic motion on GX, and one of your classmates is going to determine the equation and the graph from only the motion you create.

1)   Create your own simulation. You decide if you want to do circular or harmonic motion. Build your construction in GX. For today, all simulations will have a cycle of 6.28 seconds, and begin at the starting point of the sine function. Also, make sure it will fit on the grid below. Write an equation for the height of your point as a function of time, and sketch the corresponding graph below. (This is the answer key.)

Equation: _____

Graph:

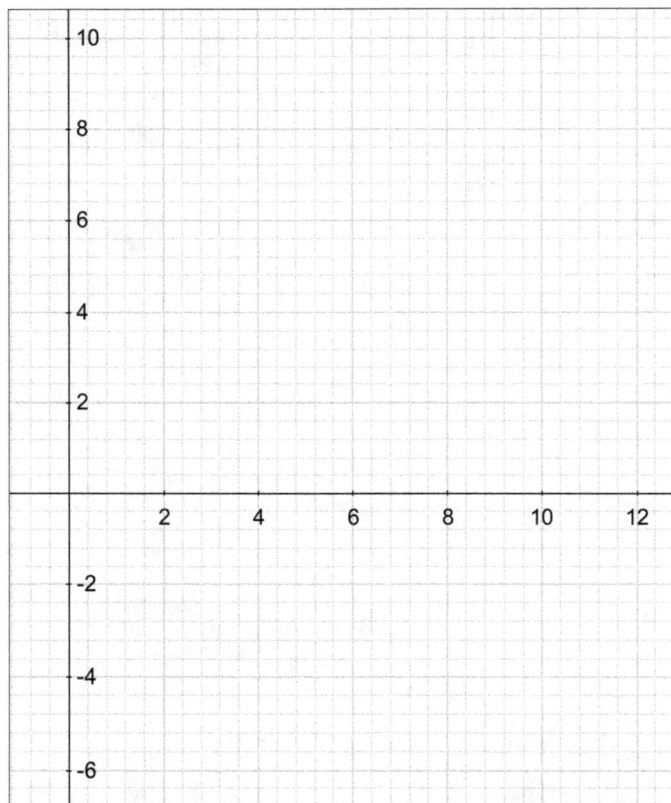

2)   Now hide all the elements of your diagram except the point that is moving. To do this, highlight each object, right click, and select Hide. All that should be visible on your screen is a coordinate grid and a point.

3) Move to a classmate's station and examine his or her model. Change the value of *t* as much as you need to by scrolling, animating, or typing in values, but do not reveal any of the constructions or constraints. Write the equation, and sketch the corresponding graph.

Equation:

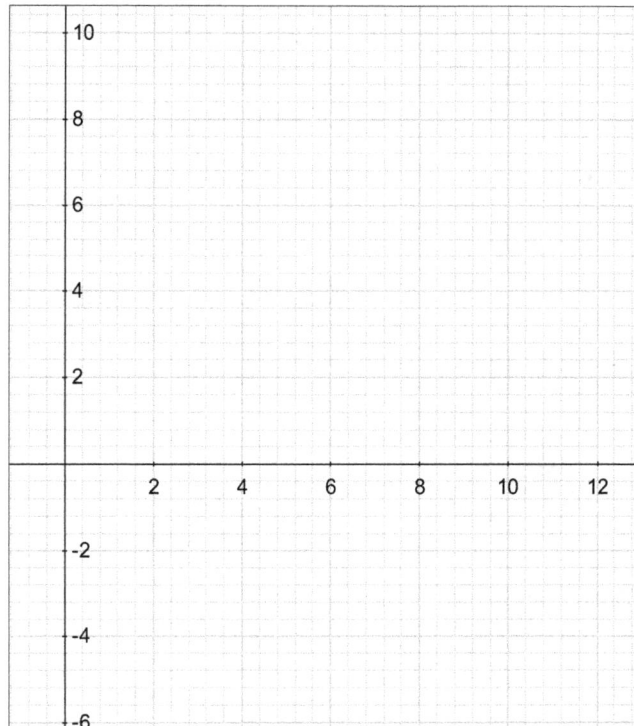

4) Repeat at a different workstation:

Equation:

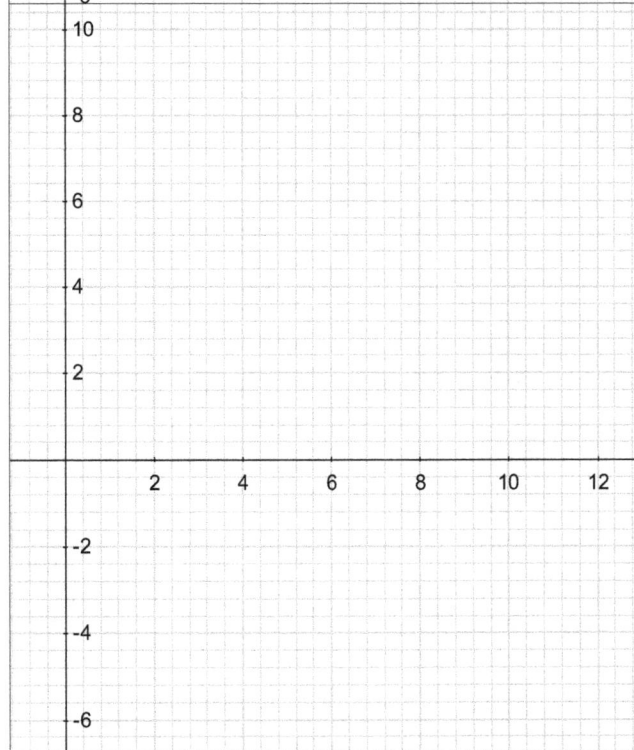

Lesson 5: Horizontal and Combined Transformations

Learning Objectives

This lesson adds horizontal translations to our previous work with vertical translations and dilations within two of the three function families we've been studying. Students will also combine all three transformations and use them in applied contexts.

Math Objectives

- Students will understand and use horizontal translations symbolically, graphically, and in contexts.

- Students will explore vertex and standard form of quadratic equations, and use technology to help them translate between them.

- Students will combine vertical translations, horizontal translations, and vertical dilations in symbolic, graphical, and contextual problems.

- Students will use vectors to perform translations.

- Students will understand the effect transformations have on asymptotes of inverse variation functions.

Technology Objectives

- Learn to use the vector and translation tools in GX.

Math Prerequisites

- Successful progress in the unit so far.

- Symbolic algebra skills in squaring binomials, distributive property, combining like terms, etc.

Technology Prerequisites

- Understanding of GX as developed in this unit so far.

Materials

- A computer with Geometry Expressions for each student or pair of students.

Overview for the Teacher

Horizontal and Combined Transformations Part A, Vertex Form

1) This problem could readily be turned into a warm up activity, depending on classroom management style. A good conversation to have is the question of <u>why</u> subtracting a number makes the graph move in a positive direction. The fact that it does is counterintuitive for most people. The first four questions are primarily designed to set up that discussion. The basic idea is that the *x* values of the solution had to increase to compensate for the 1 that was being subtracted. If students have the idea in their heads that the graph must <u>compensate</u> or in some way <u>make up for</u> the number being subtracted, they will have a more complete understanding of the mathematical pattern, and will be less likely to reverse the direction in future problems. Also, this idea is identical to vertical translations, and analogous to what happens with dilations as well, so time here is well spent.

A)   3 and –3

B)

C)   4 and –2

D)

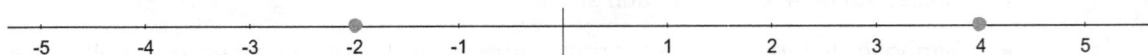

E)   It translated the graph one unit to the right.

F)   Use of GX is ok here, although point by point plotting is reasonable as well.

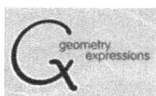

Function Transformations Lesson 5 - Horizontal & Combined
Transformations
Algebra 2; Pre-Calculus
Time required: 200 – 250 min.

$$y = x^2 - 9$$

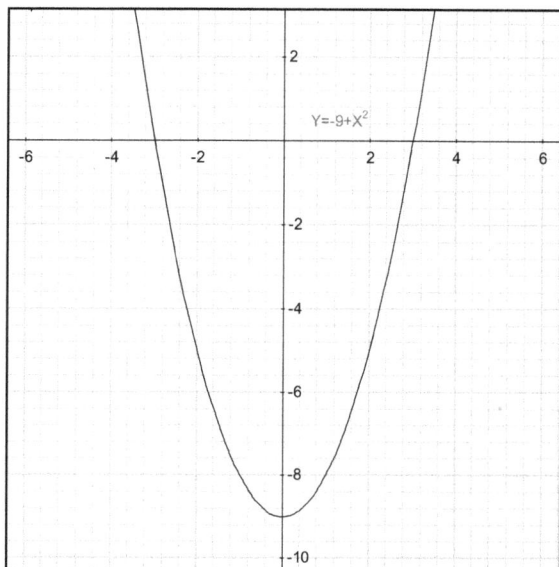

$$y = (x - 2)^2 - 9 \ [\text{or } y = x^2 - 4x - 5]$$

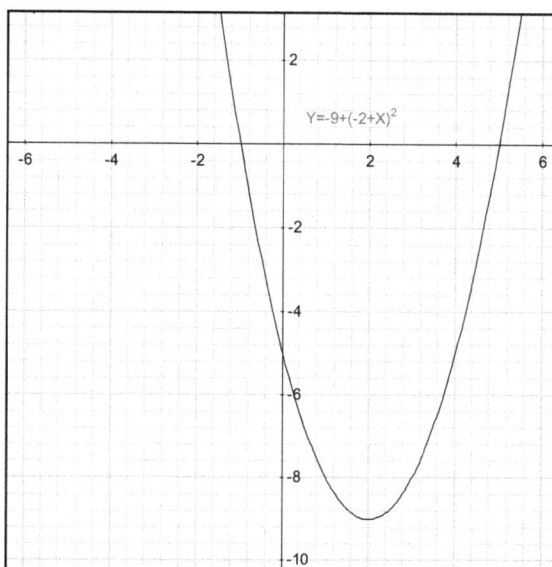

G)   Vertical translation: 9 units down

H)   Horizontal translation: 2 units to the right (This is a horizontal translation of 2,
        not a horizontal translation of –2! Watch student responses.)

I)   (2, -9) –> the vertex started at (0, 0), so the translations in G and H become
        the new coordinates.

2)   While this is key, important summary information, students shouldn't have any trouble
        producing it on their own at this point. You will probably want to restate it at later
        times in a direct instruction mode to emphasize and review.

   A)   Vertical dilation of scale factor a

   B)   Horizontal translation of h units to the right

   C)   Vertical translation of k units up

   D)   (h, k) are the coordinates of the vertex

3)   Students will probably need to be reminded that subtracting a negative is the same as
        adding a positive, so $x + 4$ is the same as $x – (-4)$. They may need a more extensive
        review of the algebra skills involved here.

A)
$$y = 5(x + 4)^2 + 7$$
$$y = 5(x^2 + 8x + 16) + 7$$
$$y = (5x^2 + 40x + 80) + 7$$
$$y = 5x^2 + 40x + 87$$

Vertex: (-4, 7)

B)
$$y = 2(x + 6)^2 - 4$$
$$y = 2(x^2 + 12x + 36) - 4$$
$$y = (2x^2 + 24x + 72) - 4$$
$$y = 2x^2 + 24x + 68$$

Vertex: (-6 , -4)

C)
$$y = 10(x - 3)^2 + 7$$
$$y = 10(x^2 - 6x + 9) + 7$$
$$y = (10x^2 - 60x + 90) + 7$$
$$y = 10x^2 - 60x + 97$$

Vertex: (3 , 7)

D)
$$y = -3(x - 5)^2 + 1$$
$$y = -3(x^2 - 10x + 25) + 1$$
$$y = (-3x^2 + 30x - 75) + 1$$
$$y = -3x^2 + 30x - 74$$

Vertex: (5 , 1)

E)   The *a* value is the same in both forms of the equation. It represents the vertical dilation scale factor.

4)   The purpose here is to introduce the idea of a vector and familiarize students with using a vector to describe a translation. If students have trouble visualizing the relationship between the vector and the location of the image, have them place the tail of the vector at the origin. They will immediately see that the head locates the vertex of the image parabola. Be careful with this demonstration, because students are prone to quickly generalize (falsely) that the "pointy end" of the vector shows the location of the vertex. It actually tells them how far and in what direction the object moved. The location of the vector is irrelevant. This may seem like splitting hairs for today's lesson, but false understandings are difficult to fix for students who move on to higher levels.

Students may need to be reminded that they can highlight a variable in the Variables tool panel, and then type in specific values at will.

A)   The image is to the left of the original (horizontal translation...)

B)   The image is to the right of the original.

C)   The image is horizontally aligned with the original.

D)   The image is below the original. (vertical translation)

E)   The image is above the original.

F)   The image is vertically aligned with the original.

5)   A)   Horizontal translation of 3, vertical translation of 4.

B)   Vertical dilation of scale factor 2

C)   Students can double-click on the equation $y = x^2$ in GX, then edit it to $y = 2 * x^2$.
        The graphs should match.

D) $y = 2(x-3)^2 + 4$

        $y = 2(x-3)^2 + 4$

E) $y = 2(x^2 - 6x + 9) + 4$

        $y = 2x^2 - 12x + 22$

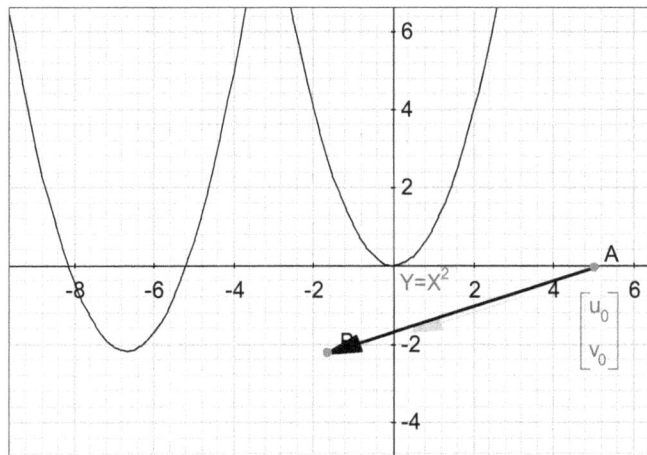

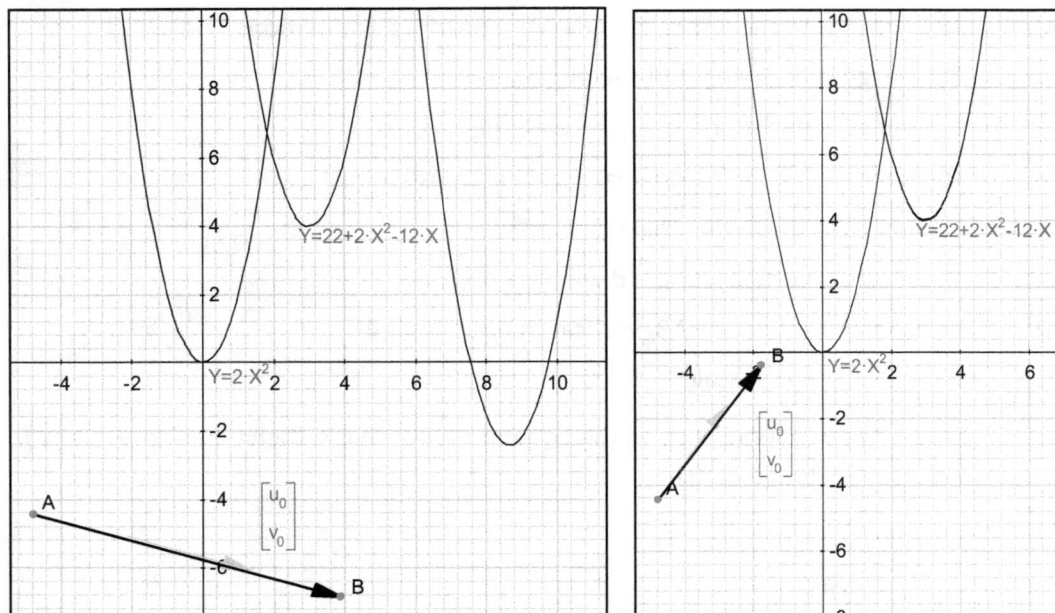

Of course, practice with symbolic techniques (i.e. completing the square) is completely appropriate here. It is simply beyond the scope of this particular lesson. Teachers are encouraged, based on time available and school priorities, to take time out of the outlined unit here to teach and have students practice the symbolic manipulation skills. As students work through the problems, they may comment on the uncertainty of how precise their answers are, or the difficulty of getting accurate values other than integers or simple fractions. These discussions could motivate the need for students to learn the more traditional techniques.

6) A) $y = 3(x - 5)^2 - 6$

Vertex: (5, -6)

$$y = 3(x - 5)^2 - 6$$
$$y = 3(x^2 - 10x + 25) - 6$$
$$y = 3x^2 - 30x + 69$$

B) $y = 4(x + 3)^2 + 5$

Vertex: (-3, 5)

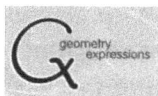

$$y = 4(x+3)^2 + 5$$
$$y = 4(x^2 + 6x + 9) + 5$$
$$y = 4x^2 + 24x + 41$$

C) $$y = -3(x+4)^2 - 5$$

Vertex: (-4, -5)

$$y = -3(x+4)^2 - 5$$
$$y = -3(x^2 + 8x + 16) - 5$$
$$y = -3x^2 - 24x - 53$$

D) $$y = 3\left(x + \tfrac{2}{3}\right)^2 - 7\tfrac{1}{3}$$

Vertex: (-2/3, -7 1/3)

$$y = 3\left(x + \tfrac{2}{3}\right)^2 - 7\tfrac{1}{3}$$
$$y = 3\left(x^2 + \tfrac{4}{3}x + \tfrac{4}{9}\right) - 7\tfrac{1}{3}$$
$$y = 3x^2 + 4x + \tfrac{4}{3} - 7\tfrac{1}{3}$$
$$y = 3x^2 + 4x - 6$$

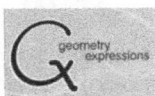

Horizontal and Combined Transformations Part B, Inverse Variation

At this point, students should be ready to combine horizontal translations with vertical translations and dilations on the other function families fairly independently. Most of this work will be done in context. Some students may need more practice with just the abstract forms, and this is easy to create. Simply give students one of three things: a) a defined set of transformations for a parent function, b) the equation or c) the graph -- have them produce the other two.

1) You may need to remind some students that a vertical line through the origin has the equation $x = 0$, horizontal line through the origin has the equation $y = 0$.

A) Student responses should look like the first figure; their computer screens should look something like the second.

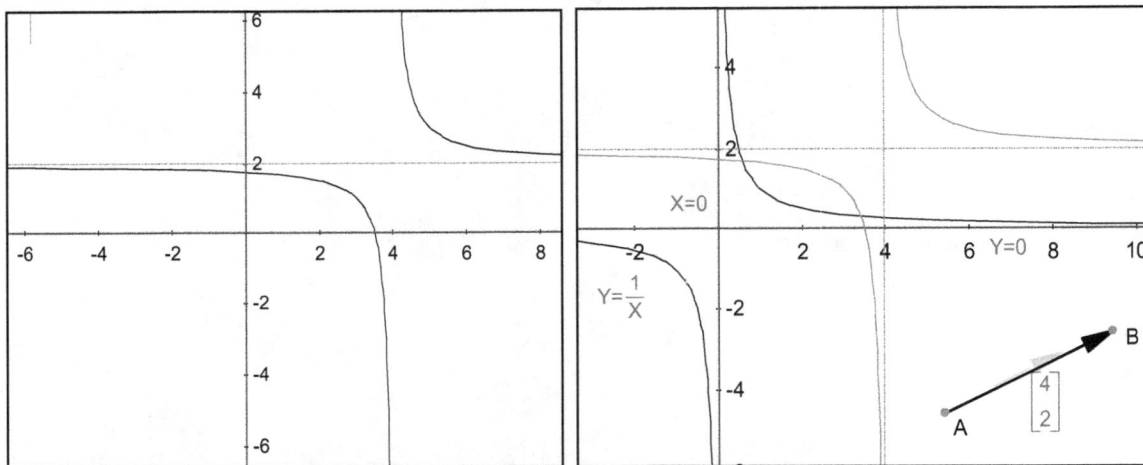

B) $y = \dfrac{1}{x - 4} + 2$

C) $y = 2; \quad x = 4$

There is an optional demonstration to illustrate the effect of the vector. It can be done as a class demo or by having students perform it.

- Fix the values of u_0 and v_0 in the variables menu.

- Select your original function, and the beginning point of

  the vector (A), and use **constrain incident** . This forces point A to stay on the function curve.

- Drag the vector around. In each position, the vector starts at a point on the original function, and points to the corresponding point after the translation.

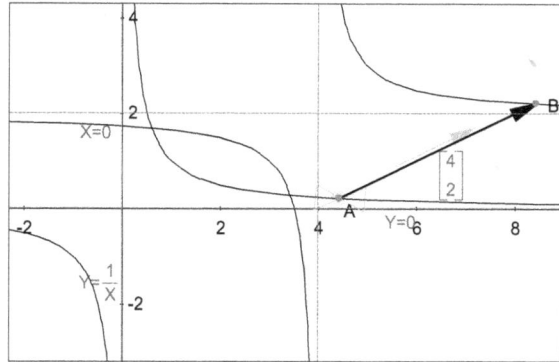

2)
    A)   Student responses should look like the first figure; their computer screens should look something like the second.

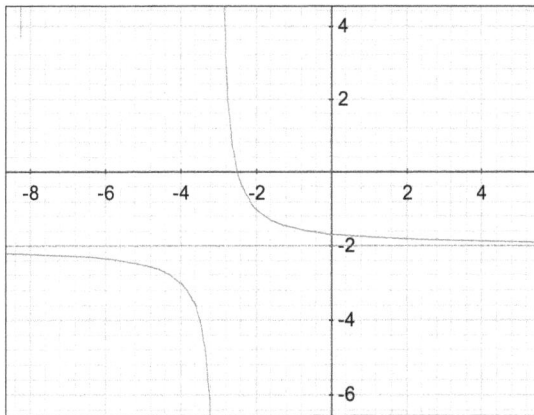

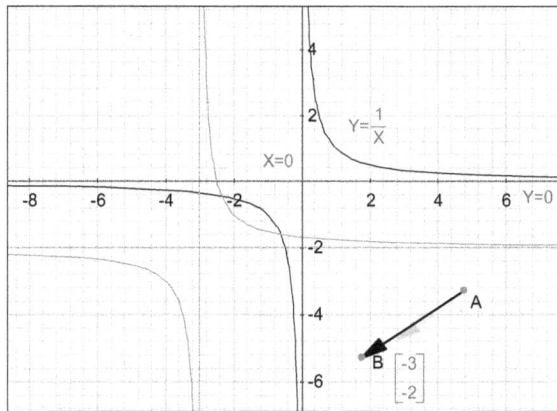

    B) $y = \dfrac{1}{x+3} - 2$

    C) $y = -2; \quad x = -3$

3)   For this problem, we ignore the second branch of the hyperbola, since negative values have no meaning in this context.

    A) $y_1 = \dfrac{500}{x}$

B)   A vertical dilation of scale factor 500.

C) $y_2 = \dfrac{500}{x-10}$ This is a good time to reinforce the idea that the function has to
compensate for the –10. Since the boss is taking 10 workers off the
manufacturing task, 10 additional workers would be required to accomplish the
same work. Thus, subtracting 10 moves the graph to the right.

D)   A vertical dilation of scale factor 500; A horizontal translation of 10 (to the right)

E) $x = 10$; $y = 0$

F)   Original Scenario: 25 workers; With car washing: 35 workers.

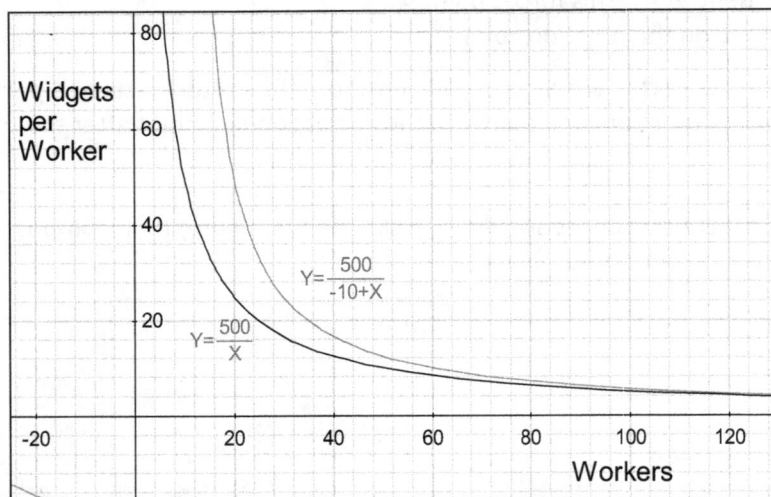

4)

A) $y = \dfrac{100}{x-4} + 1$ The scenario should be familiar enough for students to be able to
put the parameters in the appropriate places. If they struggle, suggest they use
examples that they can figure out to test their equation. For example, they can
determine that if 24 people show up, 4 will be staffing the table, leaving 20
working on clean-up. That comes out to 5 hours of work each, plus one hour for
the meeting, for a result of 6 hours to finish. If their equation gives this result,
they can be somewhat confident it is correct. Besides confirming their
equation, this is an important problem-solving strategy to model.

B)   A vertical dilation of scale factor 100; a horizontal translation of 4 units (to the
right); and a vertical translation of 1 unit (up).

C) $y = 1$; $x = 4$

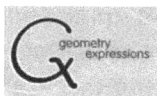

D)

Hours

$$Y = 1 + \frac{100}{-4 + X}$$

Number of Volunteers

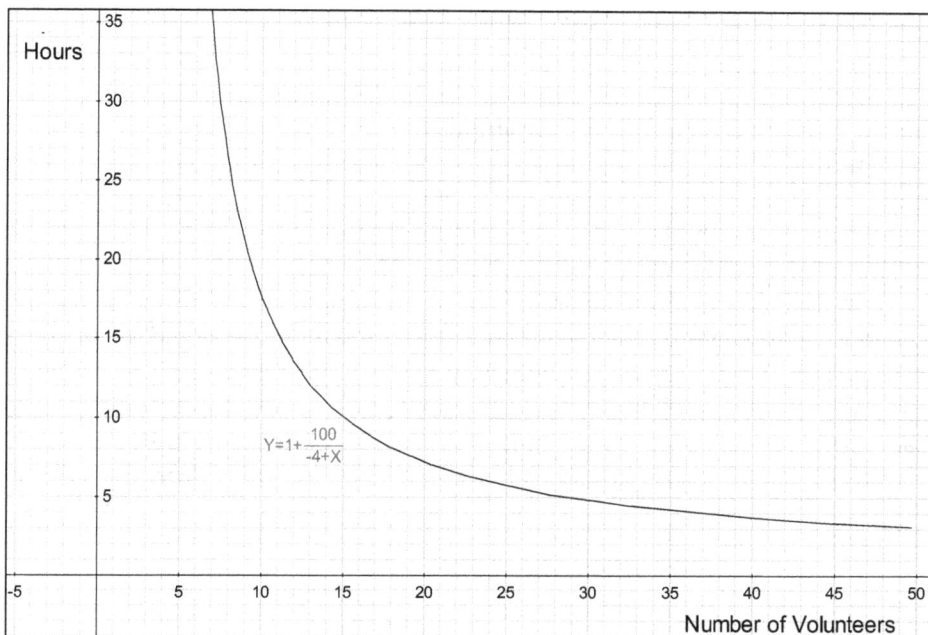

5) This problem is another model based on an inverse variation pattern. It starts with a basic pattern, which is then modified step-by-step to provide scaffolding. This should enable students to successfully figure out how the different contextual elements affect the function pattern. If they struggle, again encourage them to choose example values (like x = 10), solve the problem based on the contextual information, then solve it using their equation to check the equation's validity. If necessary remind them that there are really only three types of changes we've been studying, so there is a limited number of possibilities to check.

A) $y = \dfrac{24}{x}$ For example, 10 columns yields 2.4 cm widths.

B) $y = \dfrac{24}{x + 2}$ For example, 10 blank columns means 12 columns total, which yields 2 cm widths

C) $y = \dfrac{24}{x + 2} - 0.3$ For example, 10 blank columns means 12 columns total, which yields 2 cm widths. Then subtract 0.3 cm from each column, and the result is 1.7 cm widths.

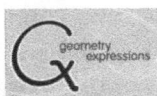

E)   Vertical dilation of scale factor 24; Vertical translation of –0.3 (shift 0.3 units down); Horizontal translation of –2 (shift 2 units to the left).

F) $x = -2$; $y = -0.3$

Students are likely to notice that their asymptotes are both negative, even though it's impossible to have a negative number of columns or a negative column width. Analysis of the intercepts should clear this up. When $x = 0$, which represents 0 usable columns, there is still a meaningful column width of $y = 11.7$ cm. This represents the width of the pre-printed columns if we let them take up the whole page. When $x = 78$, which represents a hypothetical 78 usable columns, and therefore 80 total columns, the column width before drawing in the borders would be 0.3 cm. Therefore, once we draw in the borders, there would be no space left for blank columns, or remaining width would be $y = 0$ cm. This point won't be on their pencil-and-paper graphs, but they can find it on GX. This could be an opportune moment to talk about practical vs. theoretical domain and range. In this case, both asymptotes are outside the realistic limits of the problem, but are still necessary to effectively describe the pattern.

G)

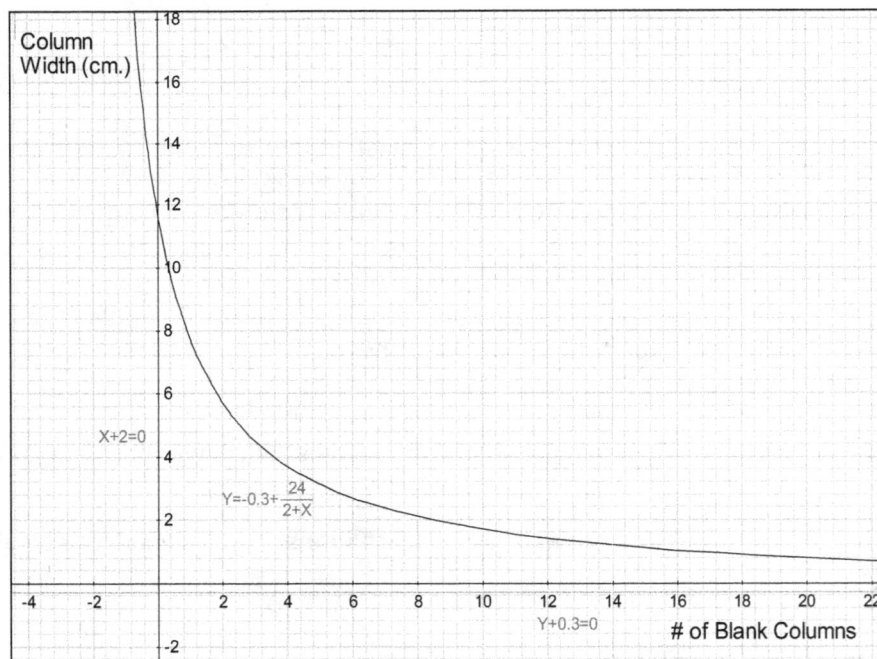

6)   Problems 6 and 7 are designed to help students practice and solidify the learning they did in #5. Some students will breeze through these as simple drills. For others, this is an

Function Transformations Lesson 5 - Horizontal & Combined
Transformations
Algebra 2; Pre-Calculus
Time required: 200 – 250 min.

important opportunity to process and truly learn the concepts that were introduced in #5. Offer less teacher-help here by referring students to #5 as a model.

A) $y = \dfrac{36}{x+1} - 0.25$

B)   A vertical dilation of scale factor 36; a horizontal translation of –1 (shift one unit to the left), a vertical translation of –0.25 (shift ¼ unit down).

C) $y = -0.25$; $x = -1$

D)

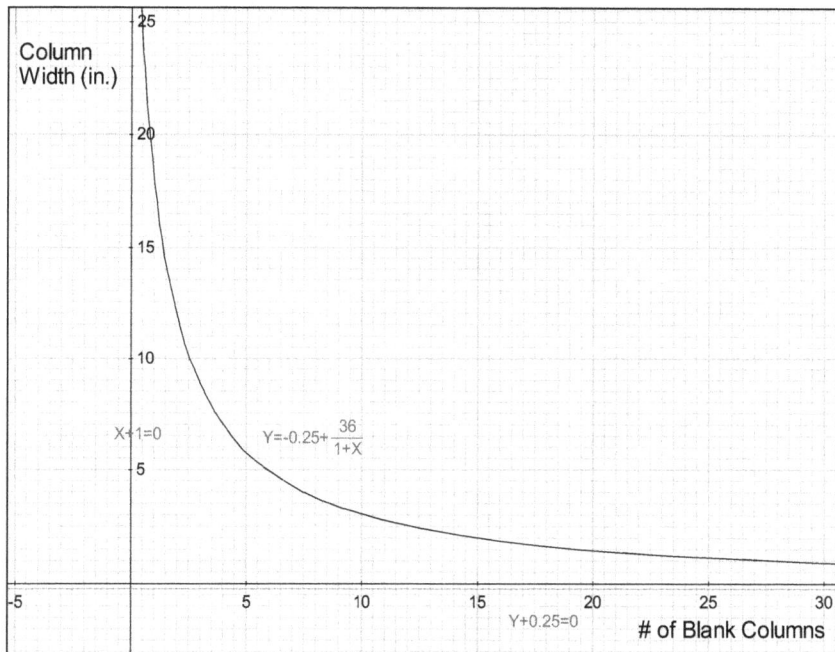

7)

A) $y = \dfrac{30}{x+3} - 0.4$

B)   Vertical dilation of scale factor 30; Horizontal translation of -3 units (shift 3 units to the left); Vertical translation of –0.4 units (shift 0.4 units down).

C) $y = -0.4$; $x = -3$

D)

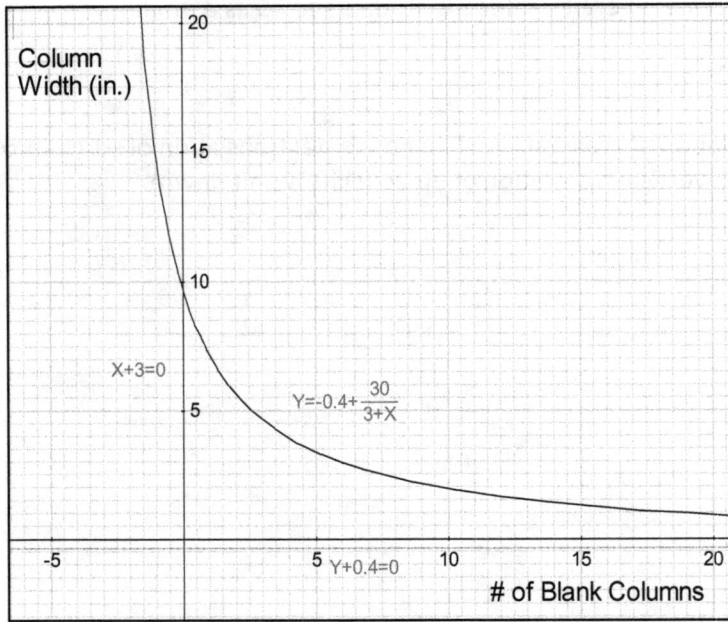

Student Worksheets

Student worksheets follow.

Name: _____

Date: _____

Horizontal and Combined Transformations Part A, Vertex Form

1)　　A)　　Let $x^2 = 9$. What does $x =$? (Two answers!)

　　　　B)　　Graph your answers to part A on a number line:

　　　　C)　　Now let $(x-1)^2 = 9$. What does $x =$?

　　　　D)　　Graph your answers to part C on a number line:

　　　　E)　　What effect did replacing x with $(x-1)$ have on your graph?

　　　　F)　　Graph the following functions

$$y = x^2 - 9$$

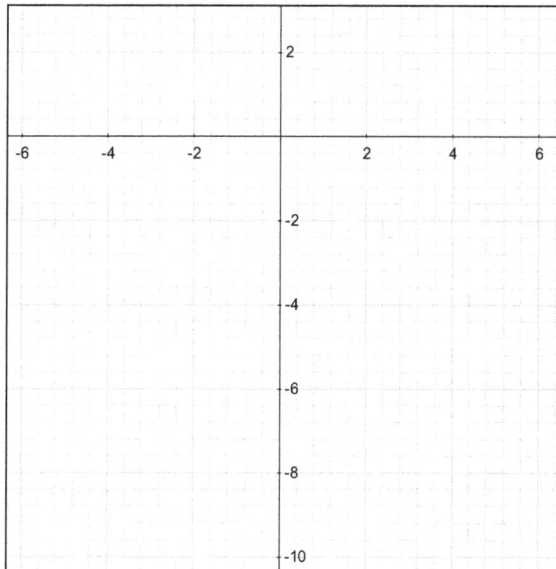

$$y = (x-2)^2 - 9 \ [\text{or } y = x^2 - 4x - 5]$$

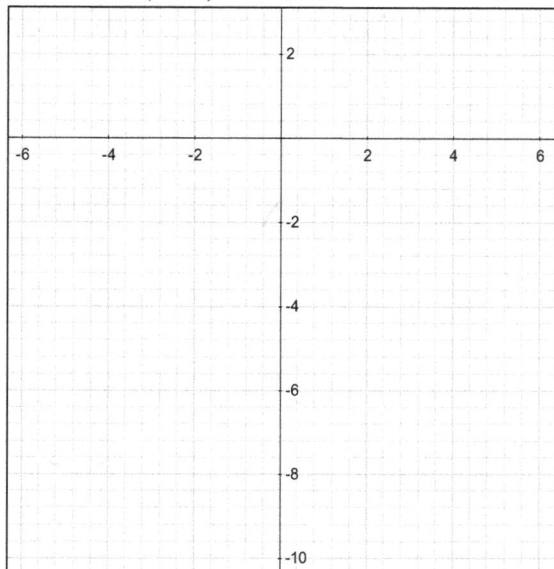

G) What does the −9 do to the parent function $y = x^2$? (review question)

H) What does the −2 do to the parent function $y = x^2$?

I) Where is the vertex of the second parabola?

Remember, replacing y with (y − k) in any equation will shift the graph k units up. Notice that replacing y with (y − k) is equivalent to adding k to the other side of the equation, which is the way it is generally written. Similarly, replacing x with (x − h) in an equation will shift the graph h units to the right. Multiplying the function by a constant still gives a vertical dilation.

2) One common form of a quadratic equation is $y = a(x - h)^2 + k$, called <u>vertex form</u>. It is simply a set of transformations done to the parent function $y = x^2$.

A) What does a do to the function?

B) What does h do to the function?

C) What does k do to the function?

D) What is the significance of the coordinates (h, k)?

3) Standard form of a quadratic equation is $y = ax^2 + bx + c$. You can convert from vertex form to standard form by using the distributive property and combining like terms. In each of the following problems, identify the vertex, then convert the equation from vertex form into standard form. Show your symbolic steps.

A) $y = 5(x+4)^2 + 7$ Vertex: (,   )

B) $y = 2(x+6)^2 - 4$ Vertex: (,   )

C) $y = 10(x-3)^2 + 7$ Vertex: (,   )

D) $y = -3(x+5)^2 - 1$ Vertex: (,   )

E) What do you notice about the a values in the standard and vertex forms? What does this value represent graphically?

A vector is a mathematical object with both magnitude and direction. It is graphically represented by an arrow. The length of the arrow represents the vector's magnitude. Velocity is one example of a physical reality that is often represented by a vector; velocity has both speed and direction. Today, we will be using a translation vector, which is used to indicate the direction and distance an object is moved.

4) Open a new file of GX. Use the function tool to draw $y = x^2$. Use **draw vector** and create a vector diagram anywhere on your page. To translate your function, highlight the parabola, use **construct translation** , and select your vector. Drag the arrow end (head) of your vector around, and note the effect on the image of your parabola.

Also highlight the whole vector and drag it around. Note that the position of the vector doesn't matter, just its length and direction. Now use **constrain coefficients** to see the horizontal and vertical components of the vector. The x coefficient, or horizontal component of the vector, is u_0, while v_0 is the y coefficient, or vertical component. Watch how the numbers in the variables window change as you drag the arrow end of the vector around.

What is true about the image parabola when ….

A)   … u_0 is negative?

B)   … u_0 is positive?

C)   … u_0 is zero?

D)   … v_0 is negative?

E)   … v_0 is positive?

F)   … v_0 is zero?

5)   Now you are going to use the translation feature of GX to help you change quadratic equations in standard form into vertex form.

Use the function tool to graph the function $y = 2x^2 - 12x + 22$ in the same file you have been using. Be sure to use * for times and ^ for exponents.

A)   Now drag the end of your vector around until the vertex of your image coincides with the vertex of the function above. What translations are required to make these graphs match? (For this example, your values will be whole numbers.)

B)   What else do you need to do to get the graphs to match?

C)   Edit the equation for the graph of $y = x^2$ on your GX page to include this dilation. Do your graphs match now?

D)   Write an equation for the function in vertex form.

E)   Check your answer for part D by using symbolic reasoning (algebra) to simplify that equation, as in problem 3. Show your steps.

6)   Use a technique similar to the one you used in problem 5 to write each function in vertex form, name the coordinates of the vertex, and then show your steps as you verify that the two equations are equivalent.

A) $y = 3x^2 - 30x + 69$

Vertex Form:

Vertex: (,)

Verify equivalence:

B) $y = 4x^2 + 24x + 41$

Vertex Form:

Vertex: (,)

Verify equivalence:

C) $y = -3x^2 - 24x - 53$

Vertex Form:

Vertex: (,)

Verify equivalence:

D) $y = 3x^2 + 4x - 6$

Vertex Form:

Vertex: (,)

Verify equivalence:

Horizontal and Combined Transformations Part B, Inverse Variation

Now that you've examined horizontal translations for functions in the family of $y = x^2$, and combined them with vertical transformations, we'll take a moment to look at another function family. Of course, the transformations work in basically the same way. However, there are different key characteristics and applications to examine.

1)   Open a new GX file and graph $y = \dfrac{1}{x}$. Create the asymptotes by using **draw infinite line**

⬜ , then **constrain implicit equation** 🖋 and typing in the equation. Repeat for the other asymptote. It is sometimes helpful to distinguish different elements in a drawing by changing line style and color. Select both of your asymptotes, right click, and select Properties. Then change the line style to one of the dashed options, and change the color to one of your choice. Now draw a vector, and constrain its coefficients to give a horizontal translation of 4 units (right) and a vertical translation of 2 units (up). Hold down the shift key to select both asymptotes and your function graph, then construct the translation.

A)   Copy the graph of the image parabola, with the asymptotes.

B)   Write the equation for the new function:

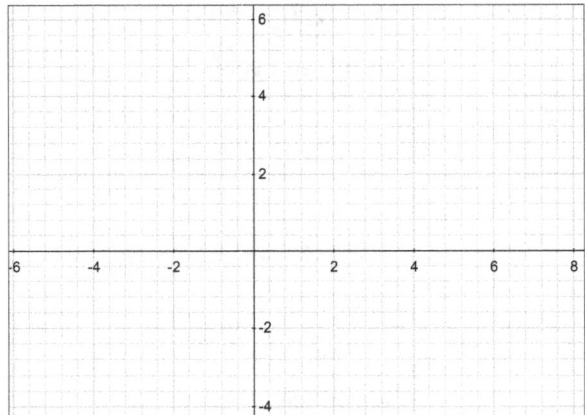

C)   Write the equations for the new asymptotes.

2) Now translate the function $y = \dfrac{1}{x}$ down two units and 3 units to the left.

A) Copy the graph of the image parabola, with the asymptotes.

B) Write the equation for the new function:

C) Write the equations for the new asymptotes.

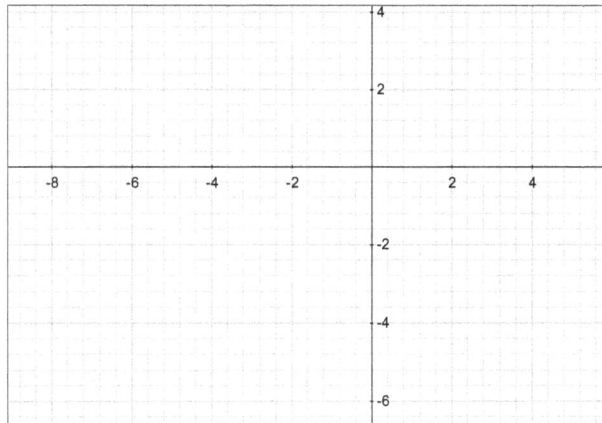

3) A manufacturing company has to produce 500 widgets in one day. The plant primarily hires people to work on a daily basis, so they never know exactly how many workers they will have on a given day. The number of widgets each worker is responsible to make depends on how many workers show up on that day.

A) Write an equation for the number of widgets each worker must make (y_1) as a function of the number of workers (*x*).

B) Describe the transformation(s) that must be made to the parent function $y = \dfrac{1}{x}$ to get this new function.

C) The boss decided this morning that he was going to use the first 10 workers to wash his collection of sports cars all day, but still expected 500 widgets. Write the new equation, using y_2 for the number of widgets.

D) Describe the transformation(s) that must be made to the parent function to get this second function.

E) Write equations for the asymptotes of the second function.

F)   Graph both functions on the axes below.

G)   Given the fact that no one can possibly make more than 20 widgets in one day, what is the minimum number of workers required for each scenario?
Original:

With the car washing:

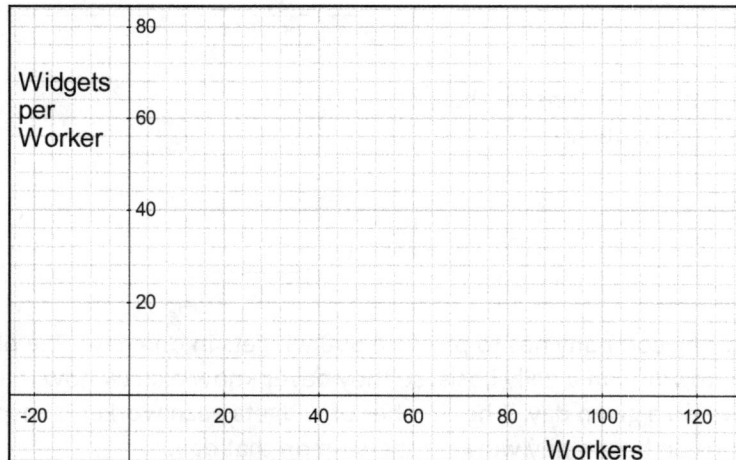

4)   An environmental awareness group is planning a riverfront clean-up event and recruiting volunteers. They estimate that there are 100 hours worth of clean-up work to do. The more volunteers who show up, the fewer hours each one has to work. Everyone must attend a one-hour meeting to go over safety, have free coffee, and hear a short speech from the mayor on the importance of what they are doing. Also, four volunteers must staff an informational table which will be set up, so those four won't be doing actual clean up.

A)   Write an equation for the number of hours it will take to complete the job as a function of the number of volunteers who show up.

B)   Describe the transformations that must be done to the parent function to produce this function.

C)   Write the equations for the asymptotes.

D)   Graph the function below.

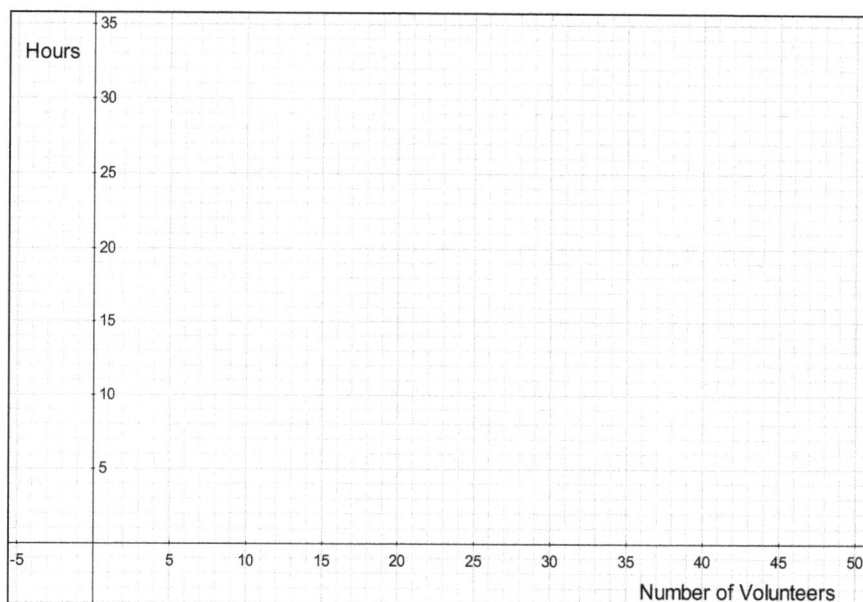

5)   Graphic Design example:
Sarah is designing a small poster-style chart, which will be written on and used to organize work tasks around her office on a weekly basis. Her boss wants a visually pleasing design, which she can print off and use for the next 1-2 years. However, her boss hasn't determined exactly how many evenly-spaced columns she wants available in the chart, but says that it will depend partially on how much space is available for each one. Part of Sarah's design task, then, is to determine column widths based on different possible numbers of columns.

A)   The width of the printable space on the poster paper they will be using is 24 cm. Write an equation for how the column width (y) depends on the number of available blank columns to be made (x).

B)   Sarah's boss now tells her that two columns will be taken up with pre-printed material. She wants those columns to be spaced equally with the blank ones. Write a new equation; x still means the number of blank columns.

C)   While working on the design, Sarah and her boss determine that the border they want between the columns will use 0.3 cm from each column. Write the new equation.

D) Describe the transformations necessary to produce your equation in part C from the parent function $y = \dfrac{1}{x}$.

E) Write equations for the asymptotes.

F) Graph the function. Include your asymptotes as dotted lines.

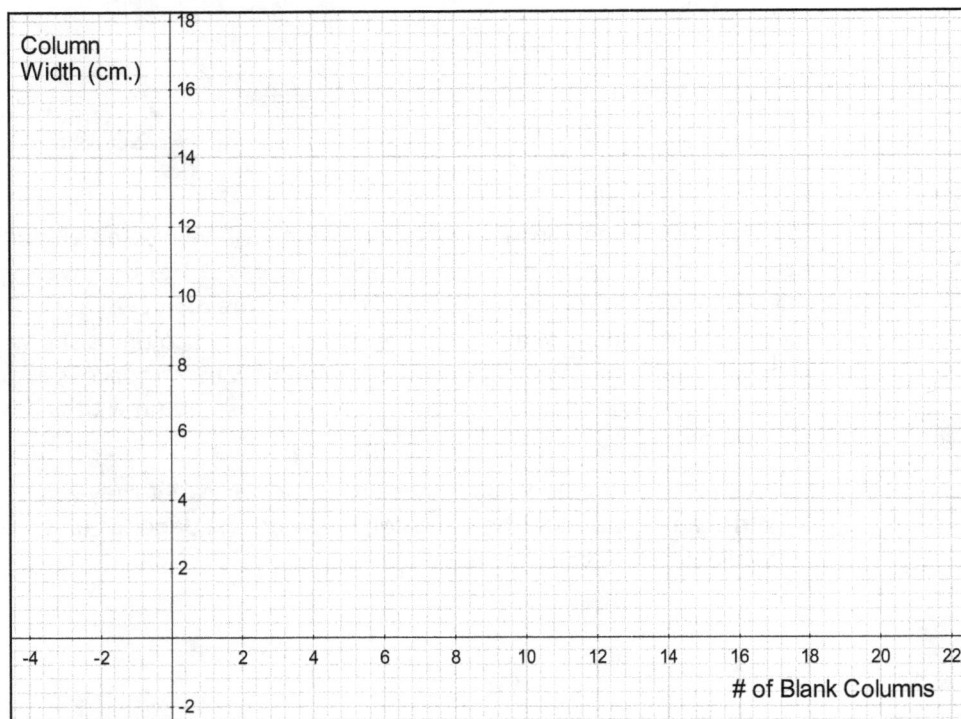

6) Sarah is called upon to make another, similar poster-chart. This time, the available paper width is 36 inches, there is only one pre-printed column, and the borders are 0.25 inch.

A) Write an equation for how the column width (y) depends on the number of available blank columns to be made (x).

B)   Describe the transformations necessary to produce your equation in part A from the parent function $y = \dfrac{1}{x}$.

C)   Write equations for the asymptotes.

D)   Graph the function. Include your asymptotes as dotted lines.

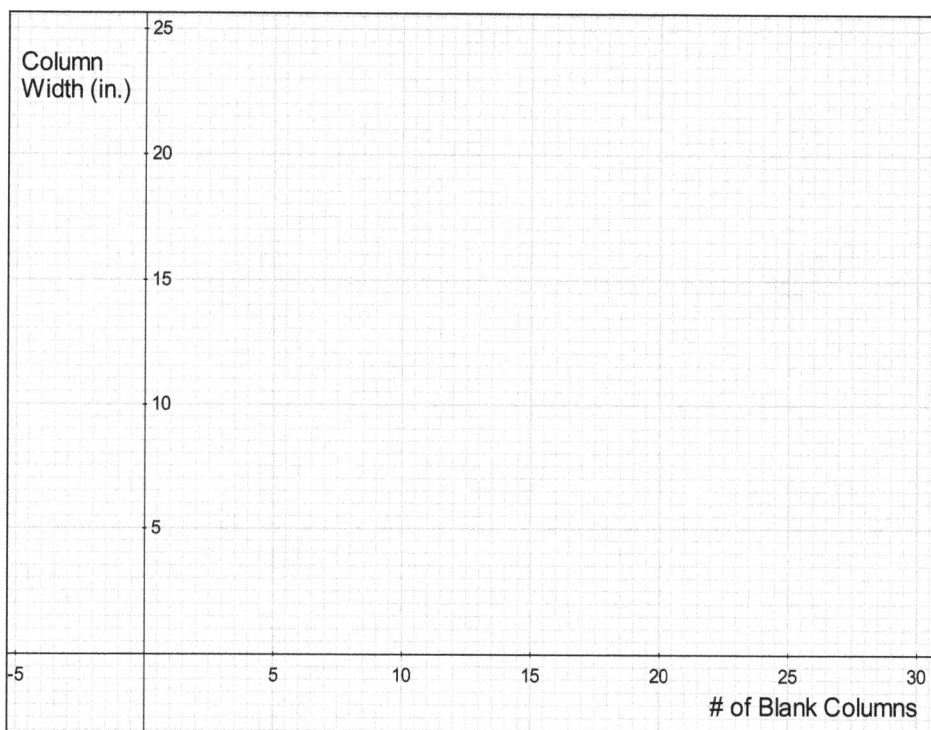

7)   Sarah is called upon to make a third poster-chart. This time, the available paper width is 30 inches, there are 3 pre-printed columns, and the borders are 0.4 inches.

A)   Write an equation for how the column width (y) depends on the number of available blank columns to be made (x).

B) Describe the transformations necessary to produce your equation in part A from the parent function $y = \dfrac{1}{x}$.

C) Write equations for the asymptotes.

D) Graph the function. Include your asymptotes as dotted lines.

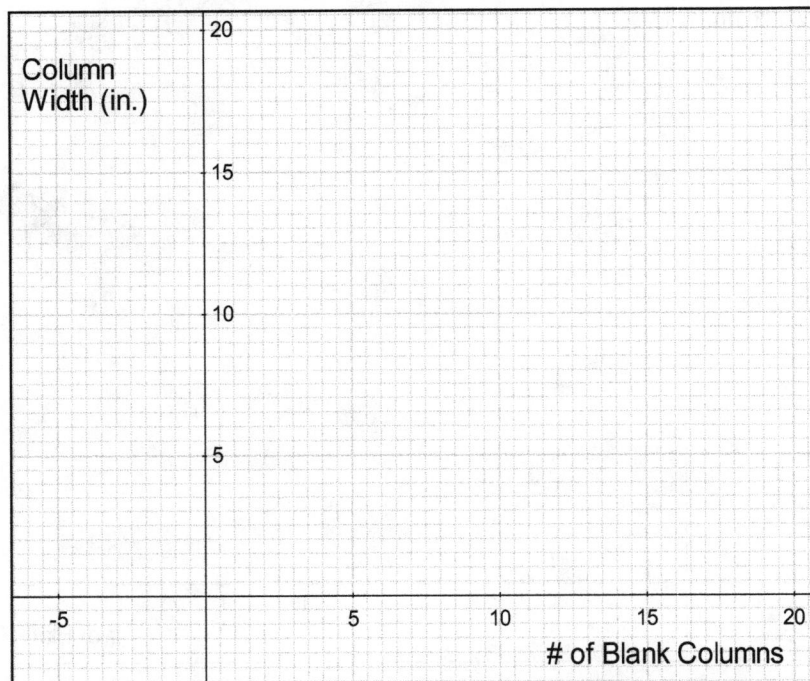

Column Width (in.)

20

15

10

5

-5

5

10

15

20

# of Blank Columns

Lesson 6: Sinusoidal Curves

Learning Objectives

This lesson completes our basic study of function transformations with the introduction and application of horizontal dilations. Of the three function families used in this unit so far, only $y = \sin(x)$ has horizontal dilations which can be definitively distinguished from vertical dilations. Horizontal translations also haven't been applied to the sine curve yet, so attention is given to those, as well as combining all four transformations. All of this will be done abstractly as well as in multiple applied contexts and simulations.

Math Objectives

- Students will understand and use horizontal dilations to influence the period of abstract and contextualized sine functions.

- Students will combine horizontal and vertical translations and dilations in sinusoidal functions.

- Students will translate between different representations of a sinusoidal pattern: symbolic, graphical, verbal/contextual, and computer simulations.

Technology Objectives

- Create simulations of circular and harmonic motion, along with corresponding function graphs.

Math Prerequisites

- Successful progress in the unit so far.
- Facility with multiplying and dividing fractions.

Technology Prerequisites

- Understanding of GX as developed in this unit so far.

Materials

- A computer with Geometry Expressions for each student or pair of students.

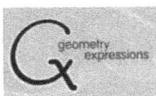

Overview for the Teacher

This lesson is designed to take several days. There aren't pre-determined breaking points, so teachers can adjust the amount of content to the schedule of their school and the abilities of their students. The extensive use of the computer modeling software should make many of the concepts more dynamic and intuitively accessible for a range of students.

Technology Notes:

The π button is located on the Symbols tool panel. Early versions of GX won't let you access the π button as you write function equations, but will let you use it in function equations if you are editing them.

Through many of the exercises, it may be helpful for students to color-code their graphs or other diagram elements on the computer. To do this, have them highlight the object, right click, and choose properties. They can also adjust line thickness, make dotted lines, etc.

1)   This problem should be routine for the students. It is mostly for the purpose of accessing prior knowledge before we move on. Double check that they are putting the correct sign on their numbers.

   A) $y = \sin(x - 2)$

   B)   Horizontal shift of 2π (left or right works.) Make sure students recognize that the decimal approximation is a multiple of π.

   C) $\sin(x) = \sin(x - 2\pi)$

   D)   Horizontal translation of $-\dfrac{\pi}{2}$ (Shift $\dfrac{\pi}{2}$ to the left.)

   E) $\cos(x) = \sin\left(x + \dfrac{\pi}{2}\right)$ or $\sin(x) = \cos\left(x - \dfrac{\pi}{2}\right)$ Students are likely to use the

   decimal approximation from the computer instead of $\dfrac{\pi}{2}$; you may need to

   clarify. Also make sure the sign, or direction of translation, is correct.

2)   This is a new application for a concept and skill students should know. It will help many students to graph these functions on the computer to visualize, even though they aren't required to.

   A) $h = 4\sin(x) + 5$

B)   Horizontal translation of $-\dfrac{3\pi}{2}$ or $\dfrac{\pi}{2}$ You may want to require students to give both answers.

C) $h = 4\sin(x + \dfrac{3\pi}{2}) + 5$ or $h = 4\sin(x - \dfrac{\pi}{2}) + 5$

3)   This problem is similar to the approach used with horizontal translations. The idea is to help students develop an understanding of why the symbolic impact of a dilation is counterintuitive. Substituting x divided by a in an equation makes the x values of the function a times greater. Again the key idea is that the x values must become greater to compensate for the division being done in the equation.

A) $x = 3$ and $x = -3$

B) $x = 6$ and $x = -6$

C) $x = 4$ and $x = -4$

D) $x = 8$ and $x = -8$

E) The graphs are being stretched horizontally by a scale factor of 2. If students don't see this right away, they will shortly.

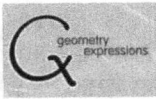

F)

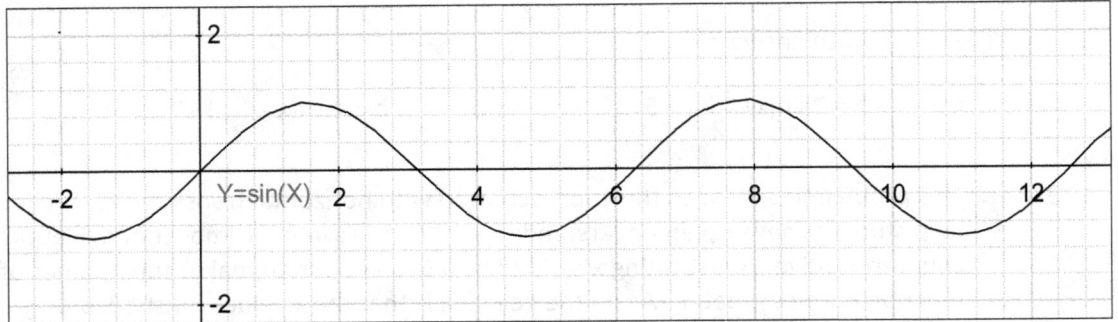

G)

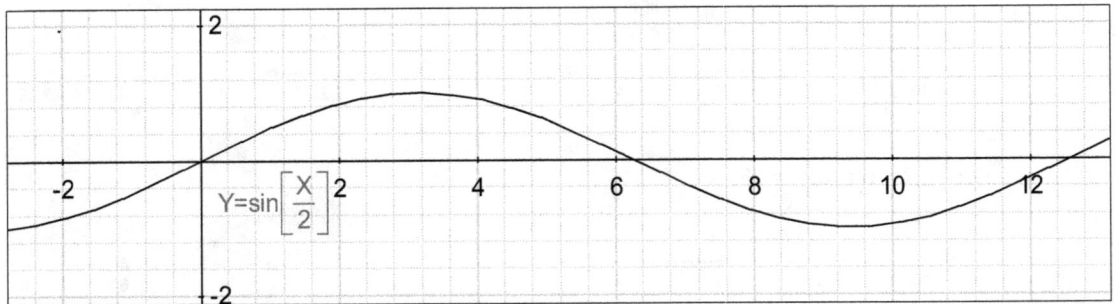

H)   The graphs are being dilated horizontally by a scale factor of 2. Specific to the sine curve: it doubles the period. (In casual student terms: They are stretched sideways to double their width.)

4)

A) $y = \sin\left(\dfrac{x}{3}\right)$

B) 6π

C)

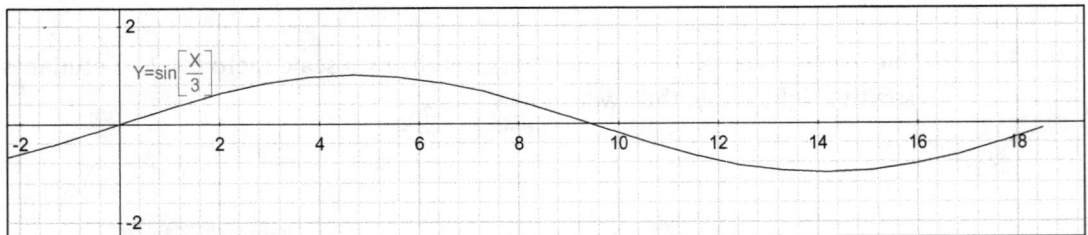

5)

A) $y = \sin(2x)$ Some students will struggle with this one because they tend to forget that dividing by ½ is the same thing as multiplying by 2. Make sure they simplify this equation, rather than leaving it as $y = \sin\left(\dfrac{x}{\frac{1}{2}}\right)$

B) π

C)

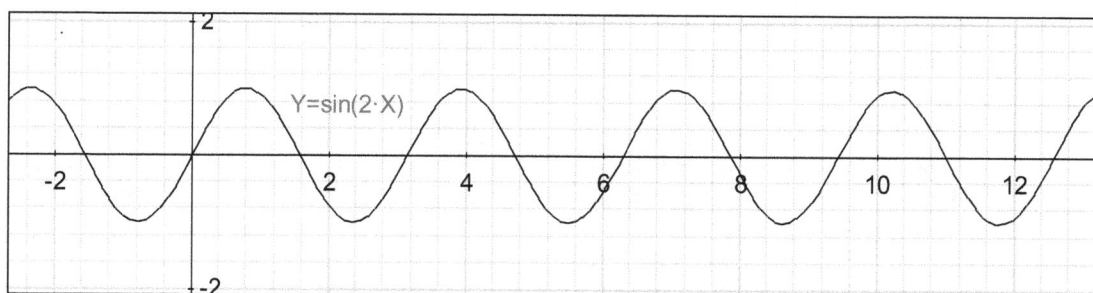

6) Students are likely to need some time and guidance on this, especially if their skill with fractions is weak.

A) A horizontal dilation of $\dfrac{1}{2\pi}$ (shrink to $\dfrac{1}{2\pi}$ or ≈ 0.159 of its former width)

B) $y = \sin(2\pi \cdot x)$

C) $y = \sin\left(\dfrac{2\pi}{5} \cdot x\right)$

D) $y = \sin\left(\dfrac{2\pi}{13} \cdot x\right)$

E) $y = \sin\left(\dfrac{2\pi}{P} \cdot x\right)$

F) This is a very important idea for students to get right. However, they shouldn't have any trouble checking the accuracy of their answer. If they ask for help, encourage them to examine their model on the software. Besides checking their answer, they can dynamically see the effect P has on the function. Below is a sample student screen when $P = 5$.

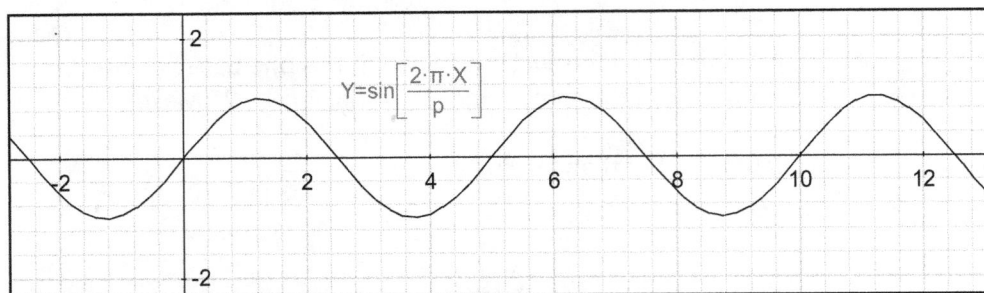

$$Y = \sin\left[\frac{2 \cdot \pi \cdot X}{p}\right]$$

7) As students start combining different transformations with sine curves, it is often helpful for them to focus on one specific key point at a time: e.g.

the natural starting point of the curve, the point where it completes one cycle, its first maximum point. Otherwise, students tend to get overwhelmed by the complexity of dealing with 4 different parameters affecting an infinitely repeating, non-linear pattern.

A) $h = \sin(\theta)$

B) $h = \sin\left(\frac{\pi}{6} t\right)$ Using different variables θ and t can help students recognize that the horizontal dilation is essentially a unit conversion. Assuming they have strong unit conversion skills, this should help conceptually.

C) $h = \sin\left(\frac{\pi}{6} t - \pi\right)$ or $h = \sin\left(\frac{\pi}{6} t + \pi\right)$ Since the phase shift is exactly half a rotation, the equation works the same way whether it is translated left or right.

D)   Option: If students adjust the number by the clock in the animation window, they can make the simulation work in real-time – i.e. a 12 second period can be reproduced in 12 seconds. The first diagram is what student screens should look like; the second is what their paper should look like.

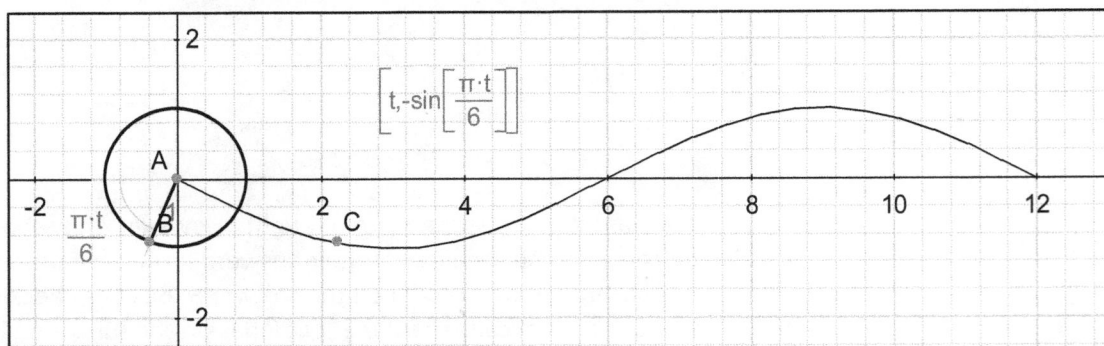

$$\left[t, -\sin\left[\frac{\pi \cdot t}{6}\right]\right]$$

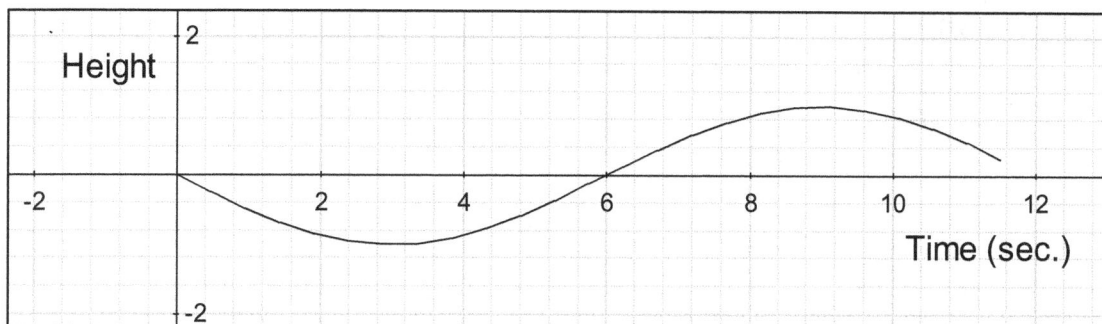

E) It is a horizontal translation of 6 seconds (units) to the right.

F) $h = \sin\left(\dfrac{\pi}{6}(t-6)\right) \Rightarrow h = \sin\left(\dfrac{\pi}{6}t - \dfrac{\pi}{6}*6\right) \Rightarrow h = \sin\left(\dfrac{\pi}{6}t - \pi\right)$ Alternately, you

could have students factor $\dfrac{\pi}{6}$ out of the equation from part C. However, unless students are particularly good at factoring, this is likely to cause students to struggle, and distract them from the main idea here. If they subtract 6 seconds from t before converting from seconds to radians, it is equivalent to subtracting π radians after the conversion. Both correspond to half a rotation.

G) Several equivalent variations are possible here. This is a good point at which to have different students produce their different equations and discuss why they are/aren't equivalent to each other and to the given problem. Moving forward, the students will want to think of horizontal translations in terms of time. Sample responses:

$h = \sin\left(\dfrac{\pi}{5}(t+2.5)\right)$ $h = \sin\left(\dfrac{2\pi}{10}(t+2.5)\right)$ --- horizontal shift is in terms

of time

$h = \sin\left(\dfrac{\pi}{5}t + \dfrac{\pi}{2}\right)$ $h = \sin\left(\dfrac{2\pi}{10}t + \dfrac{\pi}{2}\right)$ --- horizontal shift is in terms of

rotation.

$h = \sin\left(\dfrac{\pi}{5}(t-7.5)\right)$ $h = \sin\left(\dfrac{\pi}{5}t - \dfrac{3\pi}{2}\right)$

H)

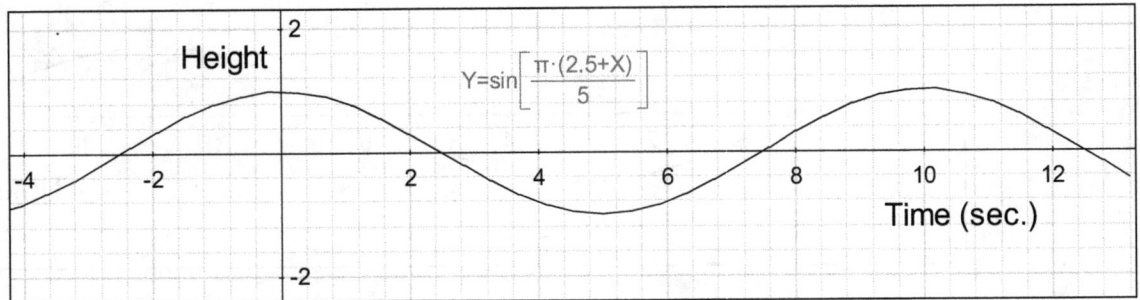

8)

A) $y = \sin\left(\dfrac{2\pi}{P} * (x - h)\right)$

B) $y = a * \sin\left(\dfrac{2\pi}{P} * (x - h)\right) + k$

C)   Allow time for students to play with the function, and encourage them to use the scroll bar. The dynamic nature of this example can help them to quickly develop an intuitive sense of sinusoidal curves and their equations.

Example student screen:

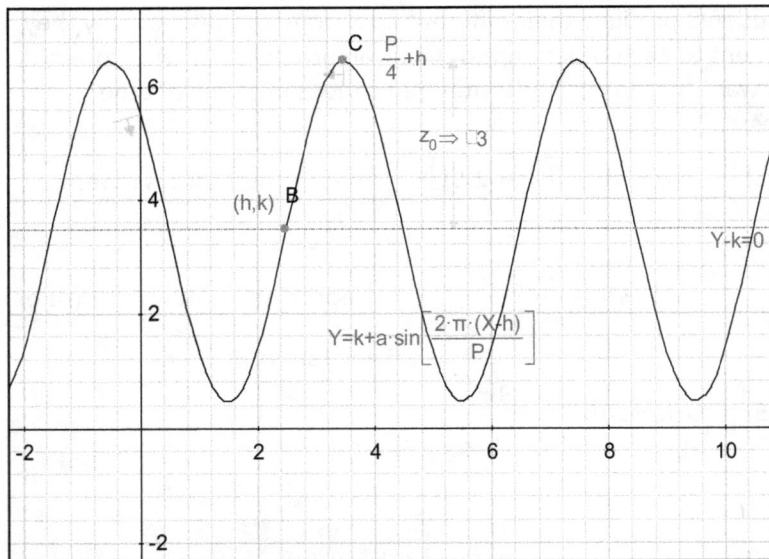

- a and P stretch the graph – vertically and horizontally

- k and h slide the graph – vertically and horizontally

- (h, k) is the image of the "starting point" of the sine curve – the point $(0,0)$ in the sine function. Students can think of this in similar terms to how they used the vertex of a parabola in relation to horizontal and vertical translations.

- This line is the central axis of the graph.

- Point C is a peak of the graph (the first peak following the translated "starting point"). You may want to help students notice that the peak of the function always happens ¼ of the way through the cycle after the natural starting point (hence the $\dfrac{P}{4}$ in the constraint.

- a

- $y = \sin(x)$

You may want to encourage students to "clean up" their screens by hiding the expressions for point C, the axis line, and the distance between them.

D)   In the equation $y = a\sin\left(\dfrac{2\pi}{P}(x-h)\right) + k$, ...

- a tells us : vertical dilation (amplitude)

- P tells us: period

- h tells us: horizontal translation (phase shift)

- k tells us: vertical translation

9)

A) $a = 35$; $h = 8$; $k = 40$; $P = 16$; Students might struggle with the h value; remind them that the wheel must make a quarter turn before it gets to the natural "starting point", and that a quarter turn takes 8 seconds. Graphically, this means a shift of 8 units to the right. Also, here and in following problems, encourage them to test their answers in the generic model on GX, and think through if the graph matches the verbal description of the problem.

B) $y = 35\sin\left(\dfrac{2\pi}{32}*(x-8)\right) + 40$ or $y = 35\sin\left(\dfrac{\pi}{16}*(x-8)\right) + 40$

C & D)   The first diagram represents students' computer screens; the second represents their papers.

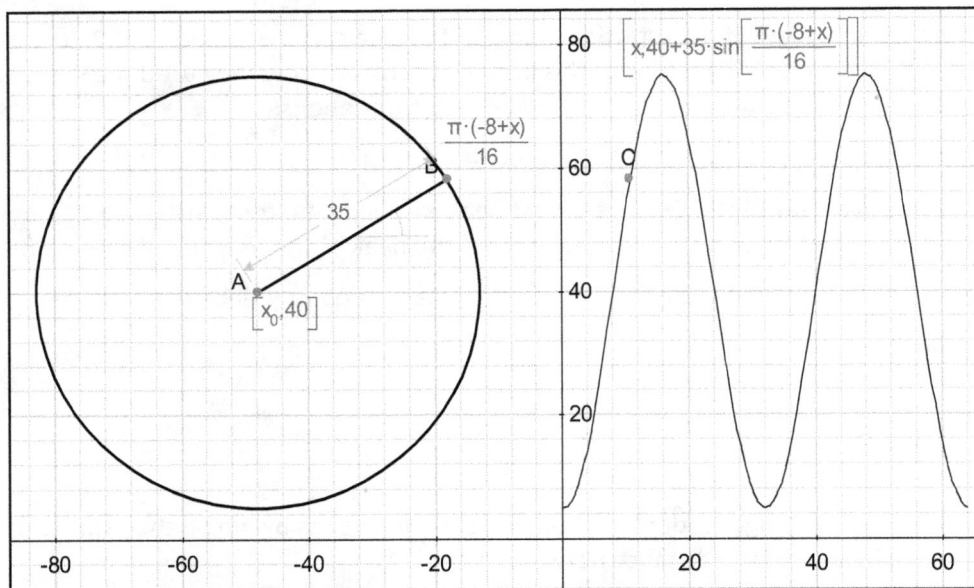

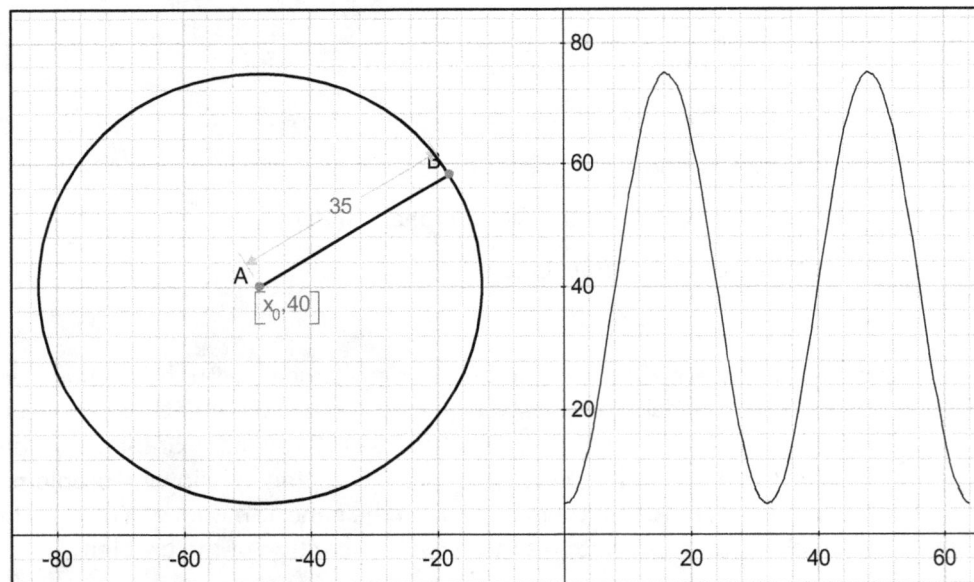

10)   The parts of question 10 can be done out of order. For example, some students will find it easier to sketch a drawing first (either by hand or by modifying their GX drawing from problem 8) and determining the transformations and equation from that. The key is for

them to see the connections between the transformations, the values of the parameters, the characteristics of the graph, and the actual applied situation. The simulations and models on GX simply give them more avenues for making those connections.

Students may object that the wheel goes through the level of the floor, but that is not unrealistic for an industrial factory.

A)

- Vertical translation of 5 (slide up 5 units)

- Vertical dilation of 7 (stretched to 7 times its regular height)

- Horizontal dilation of $\dfrac{4}{\pi} \approx 1.273$

This one is likely to throw students off. The natural period of the sine curve is approx. 6.28. We multiply that by 1.273 to get the desired period of 8 seconds. Remind them that a horizontal dilation of magnitude a is achieved by replacing x with $\dfrac{x}{a}$, or equivalently $\dfrac{1}{a}*x$. The coefficient of x in the equation will be $\dfrac{2\pi}{P}$, or in this case $\dfrac{2\pi}{8}=\dfrac{\pi}{4}$. The dilation scale factor is the reciprocal of this.

Alternately, you may want to walk students through in terms of the "unit sine curve". This creates the horizontal dilation in two steps: First dilate by $\dfrac{1}{2\pi}$, which scales the curve back to a period of one. Then dilate by 8 to get the period of 8. Combining these two produces a dilation of $\dfrac{1}{2\pi}*8=\dfrac{4}{\pi}\approx 1.273$

- Horizontal translation of -2 seconds ($-\dfrac{\pi}{2}$ radians is also acceptable) A translation of positive 6 seconds ($\dfrac{3\pi}{2}$ radians) also works.

B)   Note that the parameters are in a different order from their corresponding transformations from part A. This is done to cause students to think about the meanings and connections, rather than just copying their answers again.

- $h = -2$

- $a = 7$

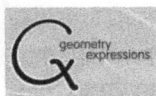

- $P = 8$
- $k = 5$

C) $\quad y = 7\sin\left(\dfrac{\pi}{4}(x+2)\right) + 5$

D & E) You may want to require teacher initials for accountability on the simulation. Students may simply modify their existing simulation model and graph to save time.

On screen:

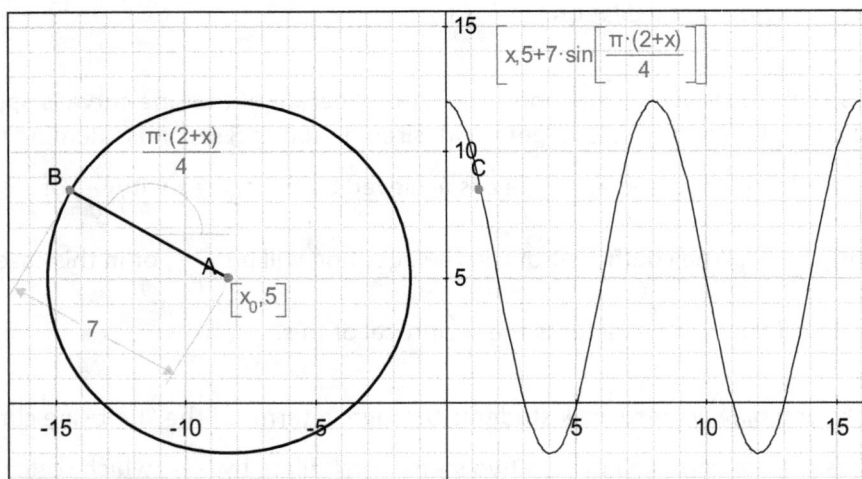

On paper:

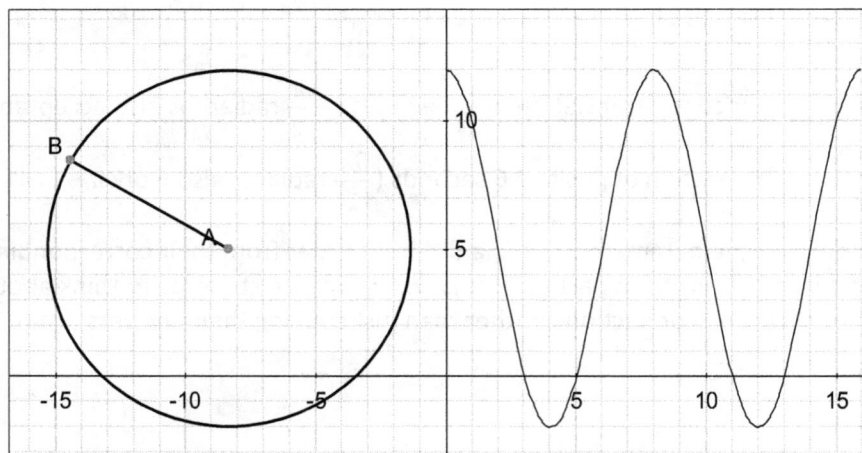

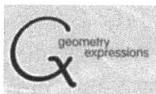

Function Transformations Lesson 6 - Sinusoidal Curves
Algebra 2; Pre-Calculus
Time required: 200 - 250 min.

11) The new issue with this problem is that students must do some interpretation of the given information to turn it into the parameters they need. If the weight goes from its lowest to its highest point in 1.5 seconds, the whole period must be 3 seconds. The vertical dilation is 0.7 – the difference between the equilibrium height and the lowest point. Also, we have a negative vertical translation since the weight is hanging below the board.

A) Vertical translation: -1.2

- Vertical dilation: 0.7

- Horizontal dilation: $\dfrac{3}{2\pi}$ -- same reasoning as problem #10

- Horizontal translation: 0.75 seconds ($\dfrac{\pi}{2}$ radians also acceptable). A translation of -2.25 seconds ($-\dfrac{3\pi}{2}$ radians) also works.

B) h = 0.75

- a = 0.7

- P = 3

- k = -1.2

C) $y = 0.7\sin\left(\dfrac{2\pi}{3}(x - 0.75)\right) - 1.2$

D & E)

On Screen:

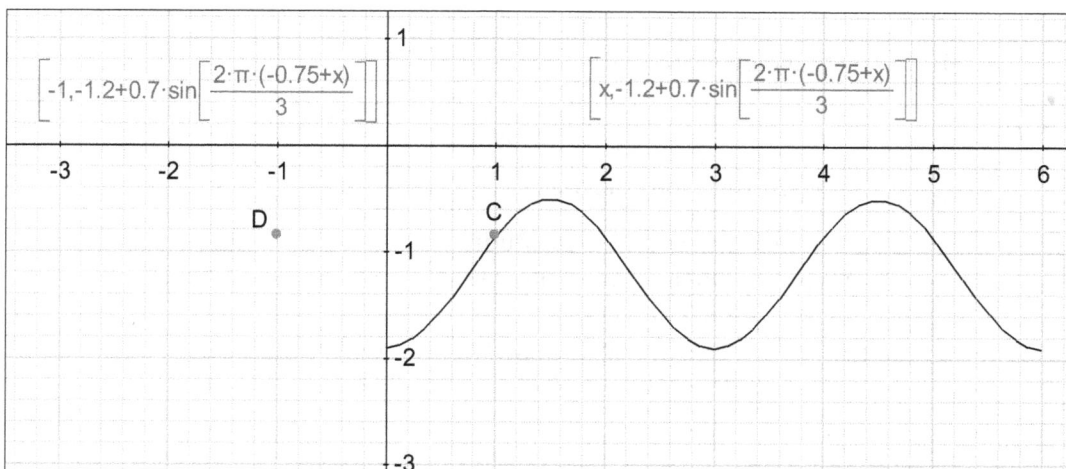

On paper:

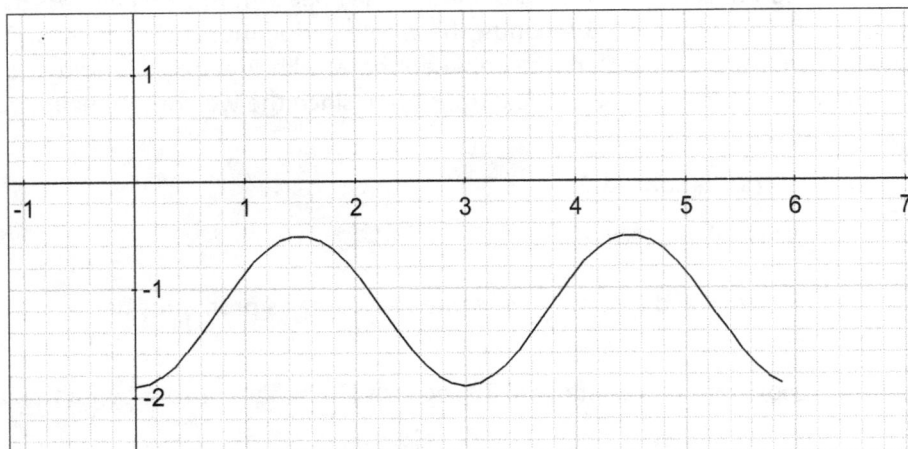

12)

A) Radius = 2 inches

B) Center height = 3 inches

C) Rotation time (period) = 10 seconds

D) Starting position: directly left of center

E) $y = 2\sin\left(\dfrac{\pi}{5}(x+5)\right)+3$ or $y = 2\sin\left(\dfrac{\pi}{5}(x-5)\right)+3$

Since the point starts half of a rotation (or 5 seconds) away from the natural starting point, the horizontal shift works identically if it goes left or right.

F) It should be a circle with the characteristics described above. The expression $\dfrac{\pi}{5}(x-5)$ or $\dfrac{\pi}{5}(x+5)$ describes the angle of rotation. To be a true real-time simulation, students should make the domain of their simulation match the number of seconds GX takes to complete it, since x = time in seconds. (E.g. if they run x from −2 to 12, they should do so in 14 seconds.)

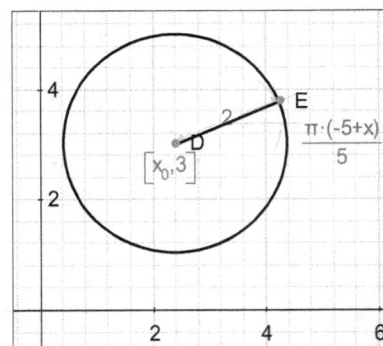

13)

A)   Radius = 0.5 feet

B)   Center height = 2 feet

C)   Rotation time (period) = 2 seconds. Note that the 2's in the expression $\frac{2\pi}{2}$ cancel.

D)   Directly to the right of center. Note that this represents no horizontal translation. Students may need to be reminded that since there is no number for h, that indicates a horizontal translation of 0 units.

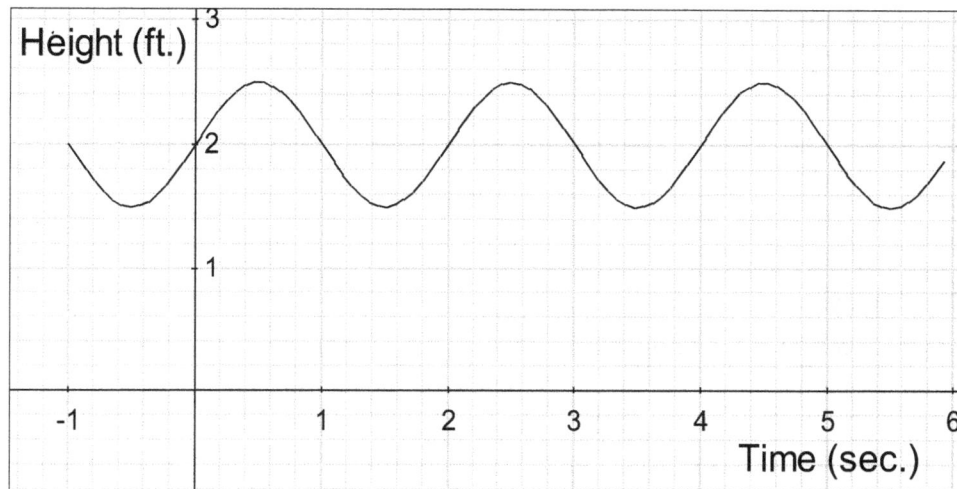

E)   Simulation should be a circle with characteristics described above. The angle of rotation is defined by $\pi \cdot x$.

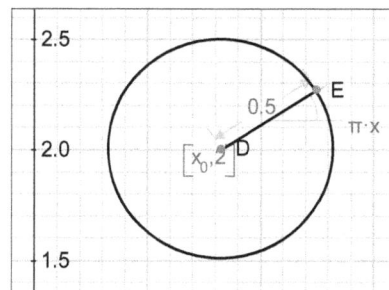

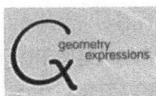

Classmate Simulation Challenge:

Obviously, answers will vary. This is the same idea and procedure as the challenge in the earlier lesson, just with more parameters to determine. This activity is optional, but may be a motivating way for students to get some practice done. It can be expanded to have students attempt more than two of their classmates' models, or possibly made into a game or contest. It can also be made more challenging by relaxing the restrictions on fitting the given graph space and starting at key points. This all depends on time, computer access, student interest, and the instructor's classroom management style.

At this point, additional practice without the computer aids and simulation may be in order. Expansion into sound waves, radio waves, daylight patterns in yearly cycles, etc. will emphasize that the skills they are practicing are good for much more than the examples given in this lesson. Such practice problems are readily available in virtually any algebra 2 or pre-calculus book, and so are not repeated here.

Student Worksheets
Student worksheets follow

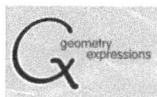

Name: _____

Date: _____

Sinusoidal Curves

The sine curve models you looked at a couple of days ago were simplified in that every example took exactly 2π seconds to run a complete cycle, and they all started at the positive x-axis or the equivalent. In this lesson, you will learn to change the starting position of a model as well as the time it takes to complete a cycle (the period).

1) Open a new GX file, and draw a graph of y = sin(x). Make sure your computer is set in radian mode: Edit/Preferences/Math/Math/Angle Mode/Radians.

A) Based on what you already know, what equation would shift this graph 2 units to the right?

Graph the function you wrote in part A. Order of operations and parenthesis are important throughout this lesson. Now check your answer by doing the translation on GX: Draw a vector

and constrain its coefficients to $\begin{pmatrix} u_0 \\ 0 \end{pmatrix}$ to indicate a horizontal translation. Select the first

function, construct translation, and then select your vector. Drag the end point of the vector until the image matches your second function graph. Did the translation match up?

B) Drag the end of the vector until the sine curve matches itself again. What translation does this? Is there a special significance to this number?

C) Write an equation reflecting this fact: sin(x) = sin(   )

D) Now edit the function equation from the graph in part A to be y = cos(x). You'll notice that it's the same basic size and shape as a sine curve. Use your vector to determine the translation necessary to turn a sine graph into a cosine graph. What is the translation?

E) Write an equation representing this relationship.

2) Consider a gear wheel that is part of an industrial machine. It has a radius of 4 feet, and is mounted 5 feet above the floor. For now, assume a period of 2π (or apprx. 6.28) seconds, a starting point directly right of center, and a counter-clockwise rotation. We want to know the height of a particular point on the wheel as a function of time.

A) Write an equation for height as a function of time. Graph it in a new GX file.

B) What translation would we have to make to the graph if we wanted time to start when the point is at its lowest height? (There are at least two possibilities.)

C) Write a new equation for this model, with the point starting at its lowest height.

3)

A) Let $x^2 = 9$. Solve for x and graph your answers on the number line.

| -10 | -8 | -6 | -4 | -2 | | 2 | 4 | 6 | 8 | 10 |

B) Now let $\left(\dfrac{x}{2}\right)^2 = 9$. Solve for x and graph your answers on the number line.

| -10 | -8 | -6 | -4 | -2 | | 2 | 4 | 6 | 8 | 10 |

C) Let $x^2 = 16$. Solve for x and graph your answers on the number line.

| -10 | -8 | -6 | -4 | -2 | | 2 | 4 | 6 | 8 | 10 |

D) Now let $\left(\dfrac{x}{2}\right)^2 = 16$. Solve for x and graph your answers on the number line.

| -10 | -8 | -6 | -4 | -2 | | 2 | 4 | 6 | 8 | 10 |

E)	What effect did replacing x with $\dfrac{x}{2}$ have on your graphs in these two examples?

F)	Graph $y = \sin(x)$

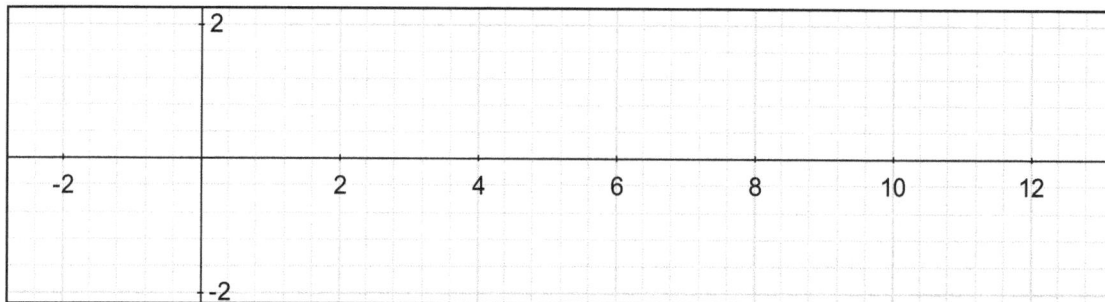

G)	Graph $y = \sin\left(\dfrac{x}{2}\right)$

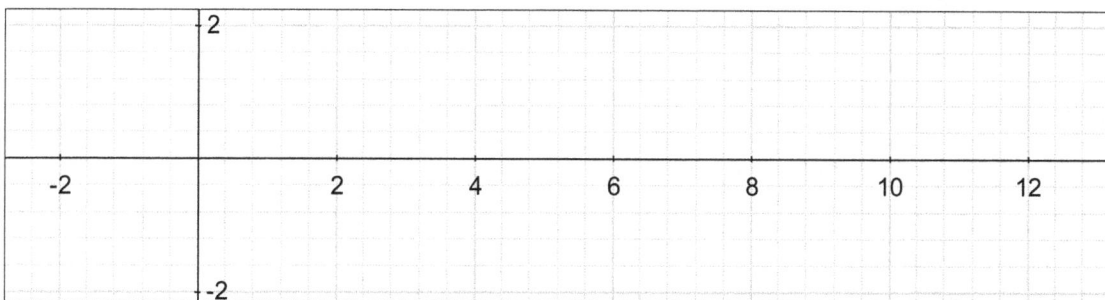

H)	What effect did replacing x with $\dfrac{x}{2}$ have on your graphs in this last example?

Remember that if you wanted to take a function $y = f(x)$ and stretch it vertically by a scale factor of a, you produced $y = a*f(x)$. Notice that this is equivalent to replacing y with $\dfrac{y}{a}$ and producing $\dfrac{y}{a} = f(x)$. For a horizontal dilation, we do the same thing with x. Replacing x with $\dfrac{x}{a}$ in an equation stretches the graph horizontally by a scale factor of a. In both cases, if $a > 1$, the graph stretches, and if $a < 1$, the graph shrinks.

4)

A)   Write an equation for a sinusoidal curve which has been stretched to triple its original width.

B)   What is the period of this function?

C)   Sketch a graph of the function below:

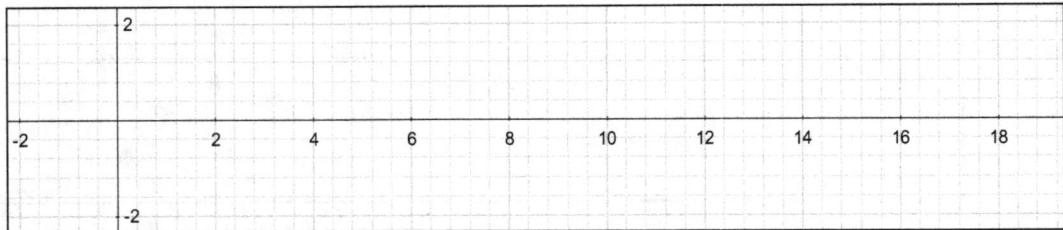

5)

A)   Write an equation for a sinusoidal curve which has been shrunk to one-half its original width.

B)   What is the period of this function?

C)   Sketch a graph of the function below:

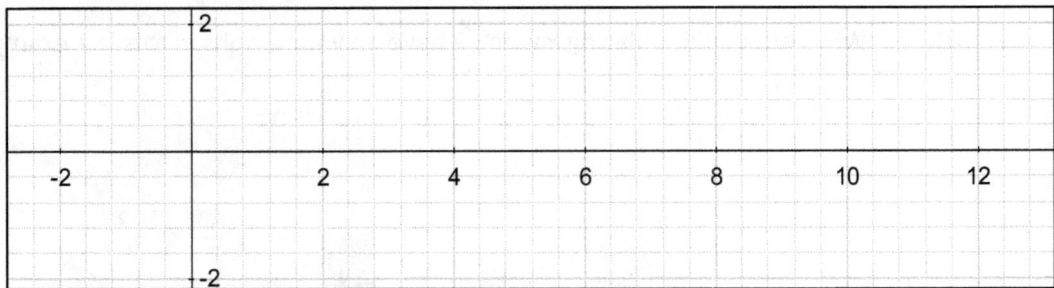

6)   Of course, in real applications of the sine curve, a complete cycle rarely happens in even multiples or fractions of π seconds. We need to be able to adjust the period.

A)   What transformation would be necessary to get a sine curve with a period of 1?

B)   Write an equation for a sine curve with a period of 1.

C)   You've now produced what you could think of as a "unit sine curve," although that's not a technical term. As you saw above, to horizontally stretch this, you divide x by the desired scale factor. Write the equation that gives a sine curve with a period of 5.

D)   Write the equation that gives a sine curve with a period of 13.

E)   Write a general equation for a sine curve with period P.

F)   Use your function tool to graph this general equation in GX. Modify the values of P to test your answer.

7)   In our previous circular motion examples, we used a period of 2π, which unrealistically made time equal to the angle of rotation. It's time to fix that. We'll start with the rotation of a unit circle.

A)   Write an equation for the height of a point rotating on a unit circle as a function of the angle it has rotated through. (This is easy – don't over-think it.)

B)   We want this circle to rotate once every 12 seconds. Modify your equation for the "unit sine curve" to reflect this.

What you are essentially doing with this horizontal dilation is converting between an angle of rotation, θ, and a time variable, t. In this case, a θ value of 2π radians equals a t value of 12 seconds. This results in the basic conversion factor of $\theta = \dfrac{2\pi}{12}t$.

C)   Modify the equation from part B to indicate a point starting to the left of center, still rotating counter clockwise. (Similar to problem #2, but this represents a half-turn rotation on the unit circle or a π translation of the function graph.)

D)   Create a simple simulation of this motion, with a corresponding function graph:

- Create a unit circle, with center at (0, 0) and draw a line segment for a radius.

- Drag your point B into the third quadrant, and constrain the angle formed by AB and the negative x axis to be $\dfrac{\pi}{6}*t$, which is equivalent to θ.

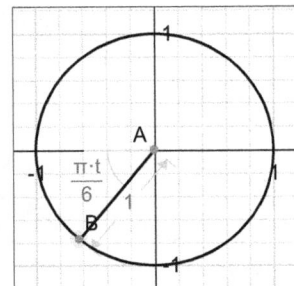

- Animate t for values from 0 to 12 to see the rotation.

- Draw point C anywhere, and constrain its x-coordinate to be t, and its y-coordinate to be the expression you wrote in part C. Make sure your expression is in terms of t, not x. GX may rewrite your expression in an equivalent form; you may ignore this. Create the locus of point C, and animate. This animation creates the function graph, and demonstrates how it is produced. Copy the graph below.

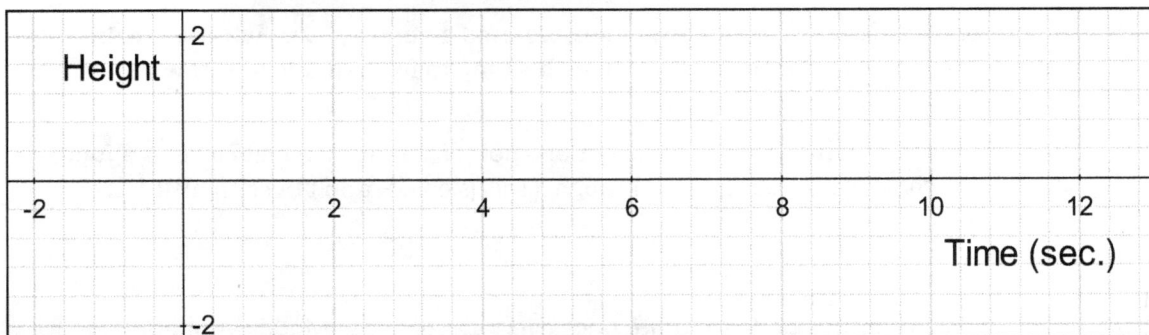

E)   Often, we will know the horizontal shift in terms of time rather than the angle of rotation. Describe the horizontal shift of the graph above in terms of time.

F) Remember that $\frac{\pi}{6}$ was the conversion factor between time and angle rotation. If we want to talk about the horizontal shift in terms of time, the shift must then be converted before the value is entered into the sine function. This creates another set of parenthesis. Show algebraically that $h = \sin\left(\frac{\pi}{6}(t-6)\right)$ is equivalent to the equation you wrote in part C.

G) Write an equation for the height of a point rotating counter-clockwise around a unit circle once every 10 seconds, starting at the top.

H) Graph the function you wrote in part G.

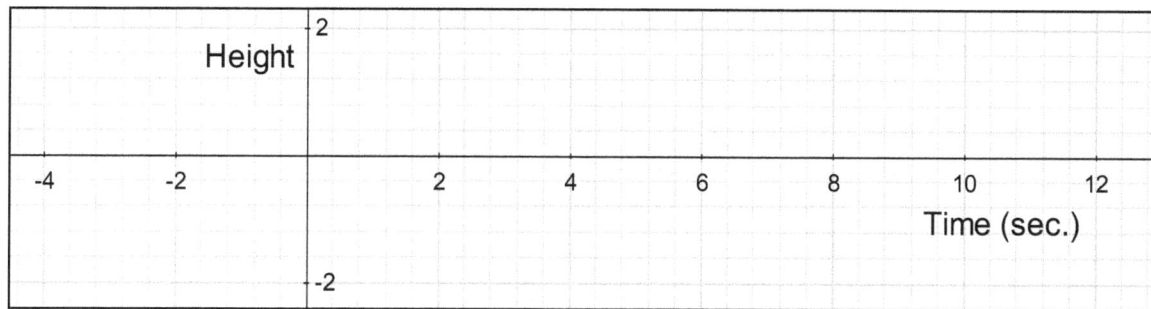

8)

A) Let h represent a horizontal shift to the right (in terms of time), and P represent the desired period of the function; write a general equation for the height of a point rotating counter-clockwise around a unit circle.

B) In addition to the parameters in part A, let k represent a vertical shift and let a represent a vertical dilation; write a general equation for any sinusoidal curve. Have your teacher check it before you move on.

Teacher Initials: _____

C)   Open a new GX file, and graph the equation you wrote in part B. Take some time to change each of the four parameters, *a, P, h*, and *k* in the variables menu and get a feel for how they change the function. You'll want to keep this file open as you move forward, as a convenient way to check equations and graphs.

- Change *a* and *P* – what do you notice?

- Change *k* and *h* – what do you notice?

Now we're going to add some elements to highlight key features of the graph.

- Draw point B not on the curve, and constrain its coordinates to (*h, k*). What is the significance of this point?

- **Draw** an **infinite line** ⬜. **Constrain** its **implicit equation** 🖋 to *y = k*. You may want to right click and change the properties to a colored, dashed line. What is the significance of this line?

- Draw point C snapped to the curve. Select the point and the curve and **constrain** it using the "**point proportional along curve**" button 🖋, set to $h + \dfrac{P}{4}$. This keeps the point on the curve, and makes its *x* coordinate ¼ of its period past *h*. What is the significance of this point?

- **Calculate** the **real distance** between point C and the line you created. Which variable does this match up with?

- Set *a* = 1, *h* = 0, *k* = 0, and *P* = 6.28. What function do you have?

D)   Summarize: in the equation $y = a\sin\left(\dfrac{2\pi}{P}(x - h)\right) + k$, ...

- *a* tells us :

- *P* tells us:

- *h* tells us:

- *k* tells us:

9)   A Ferris wheel with a radius of 35 feet is centered 40 feet above the ground. It rotates once every 32 seconds, and Joe is at the bottom of the wheel when the ride starts. You are trying to determine Joe's height above the ground after the ride has been going for *t* seconds.

A) Identify the values of the four parameters of a sinusoidal curve for this example.

$a =$

$h =$

$k =$

$P =$

B) Write the equation for the function.

C) Now open a new GX file and create a simulation of the Ferris wheel.

- Draw a circle somewhere in the second quadrant; connect points A and B.

- Constrain the radius and center to values that match the scenario. Note that the *x* coordinate of the center is irrelevant; just keep the circle to the left of the *y*-axis so it won't overlap the function graph later.

- Select segment AB and constrain its direction. Type in the expression you have in the parenthesis after sine in your equation from part B. This determines the starting point and speed of the rotation.

- Animate the simulation for two full rotations (i.e. run *x* from 0 through 64)

D) Now you are going to create the function simultaneously with the simulation.

- Draw point C anywhere, then constrain its coordinates. Type in *x* for the *x* coordinate, and your equation from B for the *y* coordinate.

- Repeat the animation. Remember that the *x*-axis represents time. Also, remember that you can adjust the "real-time" speed of your animation with the numbers next to the clock icon.

- Select C and construct its locus for the first 64 seconds of the ride. Repeat the animation.

- Now copy your simulation model and function graph below.

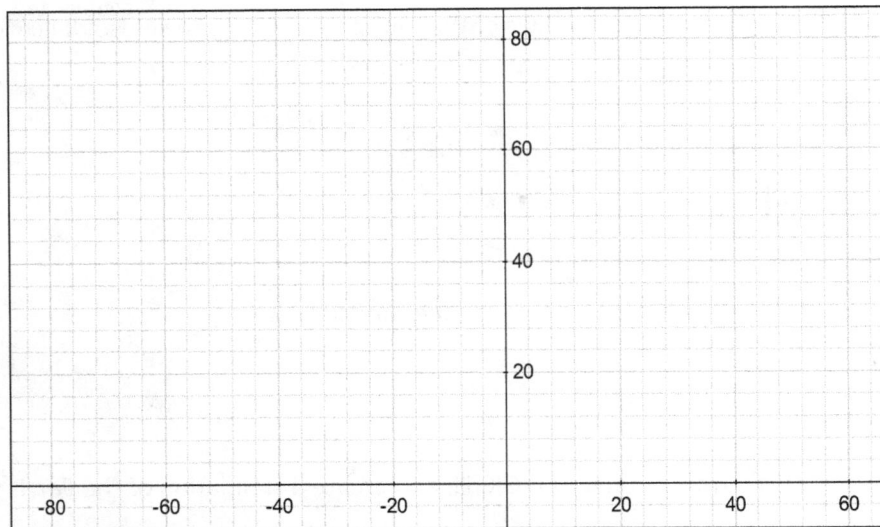

10) A gear wheel in a factory rotates counter-clockwise once every 8 seconds, and the reference point we are concerned with starts at the top. The wheel is centered 5 feet above the level of the main floor, and has a radius of 7 feet. The parts of this problem can be done out of order, depending on what is easiest for you.

A) Identify the four transformations that must be done to the parent function $y = \sin(x)$ in order to create an equation of the reference point's height as a function of time.

- Vertical Translation:

- Vertical Dilation:

- Horizontal Dilation:

- Horizontal Translation:

B) Identify the values of the four parameters of a sinusoidal curve for this example. These correspond with the answers from A, but are in a different order.

- $h =$

- $a =$

- $P =$

- $k =$

C) Write the equation for the function.

D)   Now open a new GX file and create a simulation of the gear wheel. Set the animation to run the simulation through two complete cycles.

E)   Now create the function simultaneously with the simulation, like you did in problem #9. Copy your simulation model and function graph below.

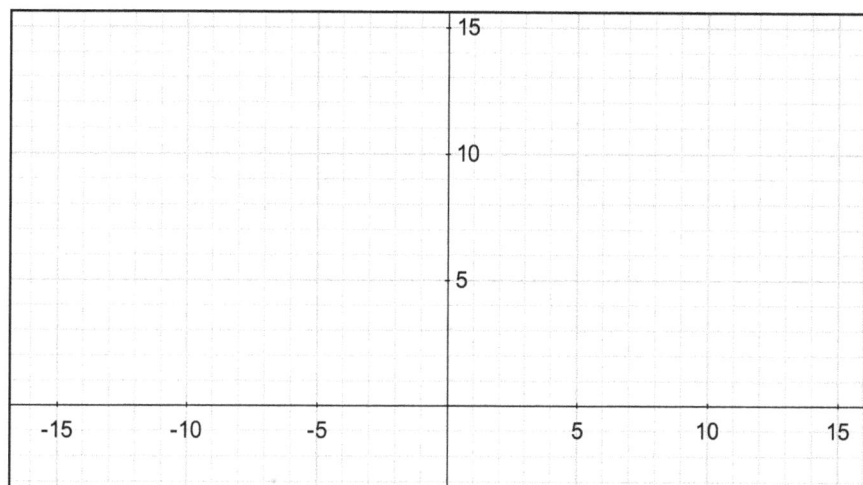

11)   A weight is hanging on a spring below a board. It is stretched and released, creating a harmonic motion pattern as the weight oscillates up and down. At equilibrium, the weight rests 1.2 meters below the board. It is released at time zero from its maximum stretch, which puts the weight 1.9 meters down. It reaches its maximum height after 1.5 seconds.

A)   Identify the four transformations that must be done to the parent function $y = \sin(x)$ in order to create a equation of the weight's height as a function of time.

- Vertical Translation:

- Vertical Dilation:

- Horizontal Dilation:

- Horizontal Translation:

B)   Identify the values of the four parameters of a sinusoidal curve for this example.

- $h =$

- $a =$

- $P =$

- $k =$

C) Write the equation for the function.

D) Open a new GX file and create a simulation of the weight. This will simply look like a point oscillating up and down. Set the animation to run the simulation through two complete cycles.

E) Now create the function simultaneously with the simulation, like you did in problem #9. Copy your function graph below.

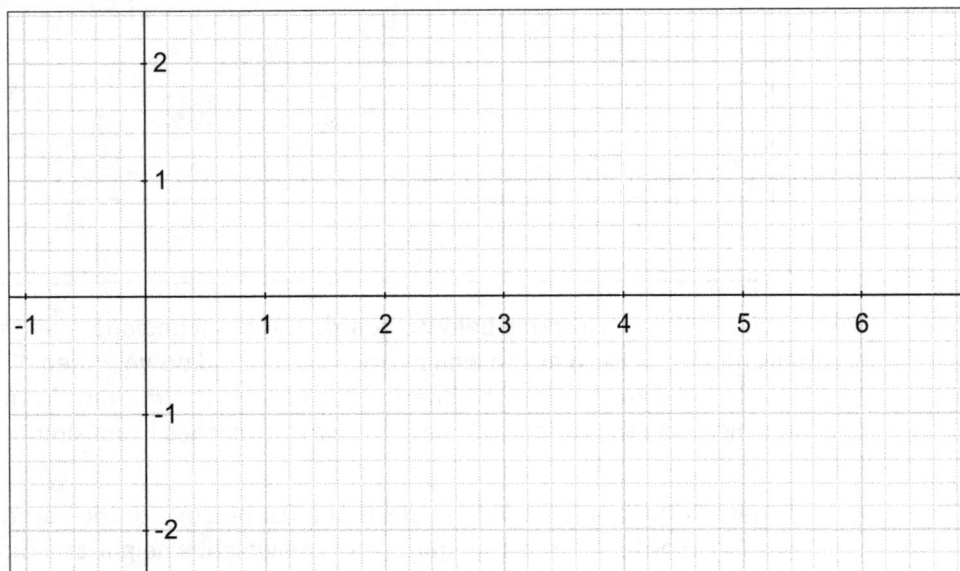

12) A graph of the height (in inches) of a point on the edge of a wheel after t seconds of counter-clockwise rotation is given below:

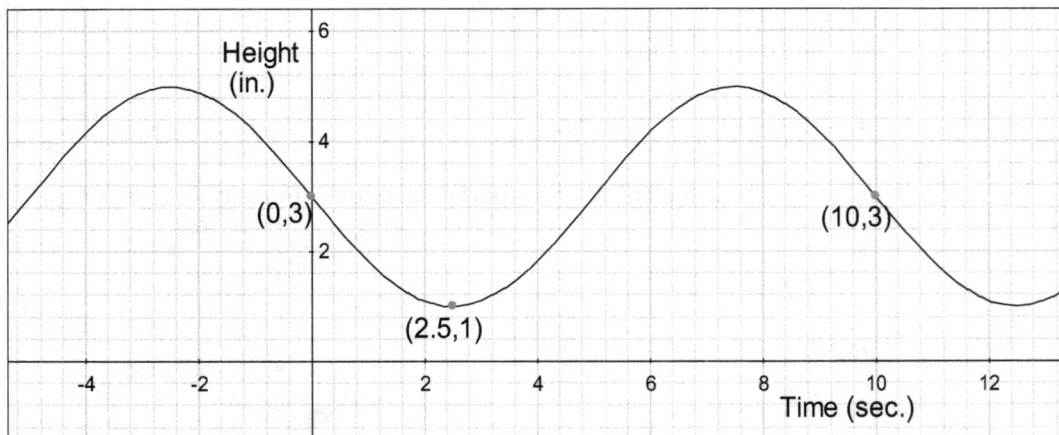

A)   What is the radius of the wheel?

B)   What is the height of the center of the wheel?

C)   How long does one complete rotation of the wheel take?

D)   At what position does the point in question start?

E)   Write an equation for the function.

F)   Make a simulation model of the wheel rotating in real-time.

Teacher Initials: _____

13)   The equation for the height (in feet) of a point on the edge of a wheel after t seconds of counter-clockwise rotation is given below:

$$y = 0.5\sin(\pi \cdot x) + 2$$

A) What is the radius of the wheel?

B) What is the height of the center of the wheel?

C) How long does one complete rotation of the wheel take?

D) At what position does the point in question start?

E) Make a graph of the function:

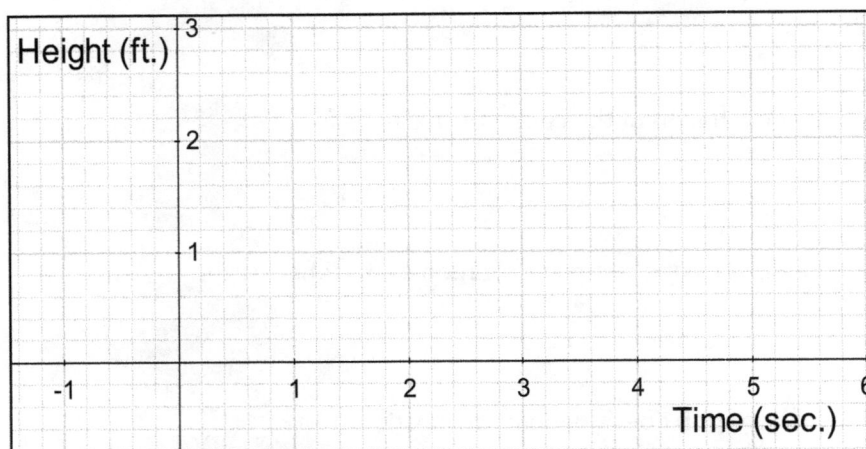

Height (ft.)

3

2

1

-1   1   2   3   4   5   6

Time (sec.)

F) Make a simulation model of the wheel rotating in real-time.

Teacher Initials: _____

Classmate Simulation Challenge

You are going to create a simulation of either circular or harmonic motion on GX, and one of your classmates is going to determine the equation and the graph from only the motion you create.

1)   Create your own simulation. You decide if you want to do circular or harmonic motion. Build your construction in GX. Make sure the function graph will fit on the grid below. Also make the starting point one of the four key points: a maximum value, middle value, or minimum value. Write an equation for the height of your point as a function of time, and sketch the corresponding graph below. Use *t* for your independent variable. (This is the answer key.)

Equation: _____

Graph:

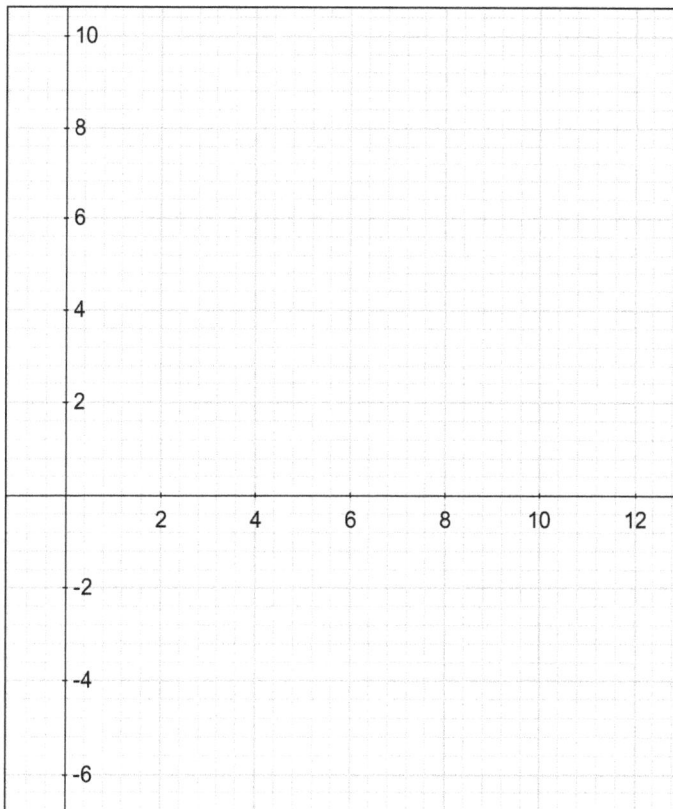

2)   Now hide all the elements of your diagram except the point that is moving. To do this, highlight each object, right click, and select Hide. All that should be visible on your screen is a coordinate grid and a point.

3) Move to a classmate's station and examine his or her model. Change the value of *t* as much as you need to by scrolling, animating, or typing in values, but do not reveal any of the constructions or constraints. Write the equation, and sketch the corresponding graph.

Equation:

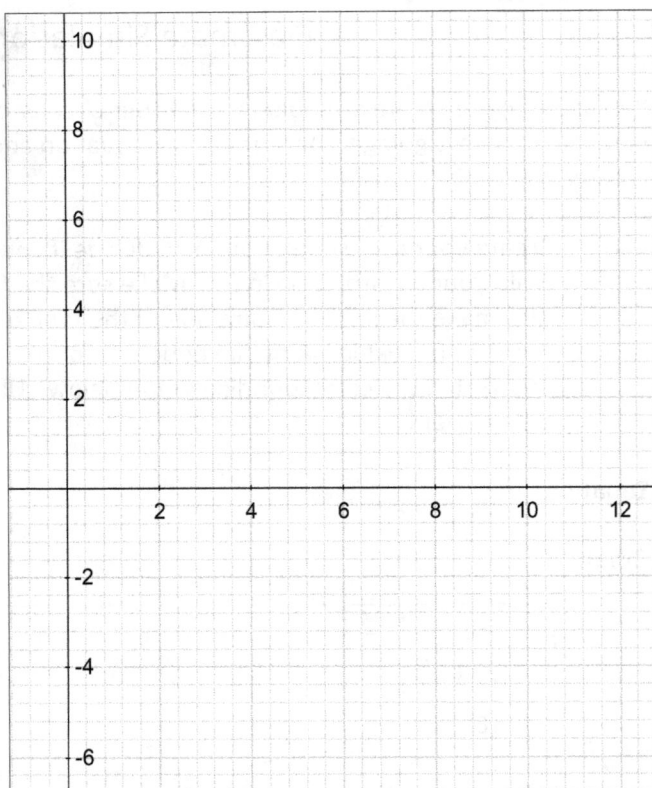

4) Repeat at a different work station:

Equation:

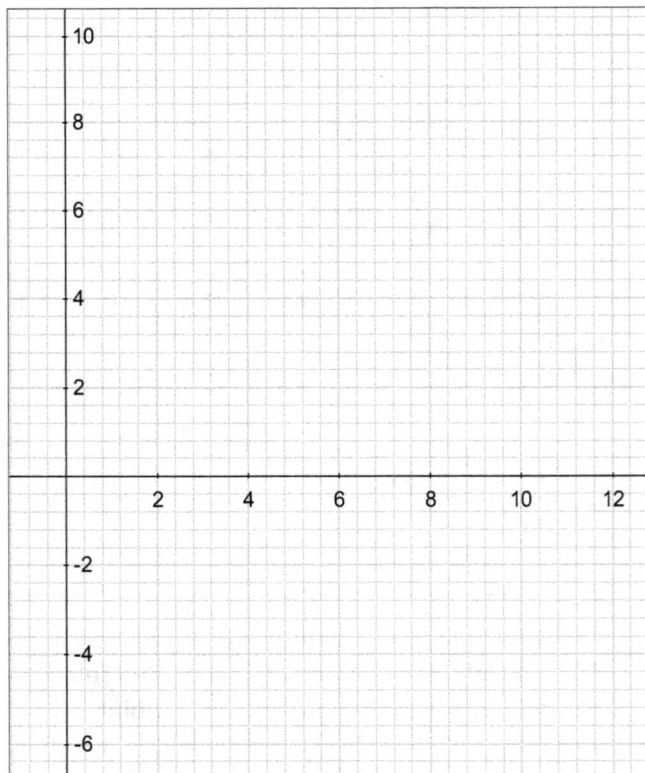

Extension A: Circles and Ellipses

Learning Objectives

Math Objectives

- Students will write the general forms of Cartesian equations for circles and ellipses, and be able to identify the key characteristics represented by each in terms of transformations.

Technology Objectives

- None

Math Prerequisites

- Successful completion of the core function transformations unit.

- Algebra manipulation skills with squares and square roots.

Technology Prerequisites

- Knowledge of Geometry Expressions as developed in the core function transformations unit.

Materials

- A computer with Geometry Expressions for each student or pair of students.

Overview for the Teacher

<u>Function Transformations Ext. A: Circles and Ellipses</u>

This short extension of the transformation units explores the use of transformation rules on graphs that are not functions. Specifically, it demonstrates the relationship between circles and ellipses as unequal horizontal and vertical dilations. Students also apply translation rules and formulate the general Cartesian equations.

Students may already be familiar with the equation for the unit circle. If that is the case, the beginning should flow quickly and be a simple review of why that formula works.

1) Students should know several ways to create the unit circle. Any of them should work. One possibility: Draw a circle anywhere, then constrain its center to (0,0) and constrain its radius to 1.

A) AC

B) BC

C) 1

D) $x^2 + y^2 = 1$

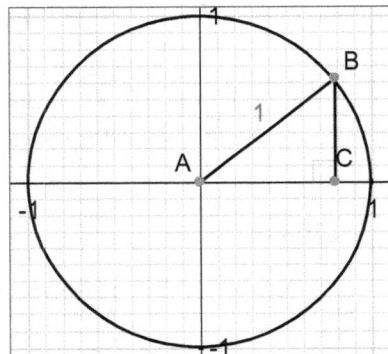

2) A) $x^2 + y^2 = 4$

B) $x^2 + y^2 = 9$

C) $x^2 + y^2 = r^2$

3) A) $(x-3)^2 + (y+2)^2 = 1$

B) $x^2 - 6x + 9 + y^2 + 4y + 4 = 1$

C) $x^2 - 6x + y^2 + 4y + 12 = 0$

D) The same translation rules apply

E) $(x-h)^2 + (y-k)^2 = r^2$ This is a good place for a checkpoint to make sure students are on the right track.

F) GX equation: $x^2 + y^2 - 2xh + h^2 - 2yk + k^2 - r^2 = 0$

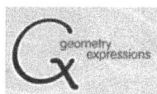

Equivalence: $\begin{cases} (x-h)^2 + (y-k)^2 = r^2 \\ x^2 - 2xh + h^2 + y^2 - 2yk + k^2 = r^2 \\ x^2 + y^2 - 2xh + h^2 - 2yk + k^2 - r^2 = 0 \end{cases}$

4) A) $\left(\dfrac{x}{a}\right)^2 + \left(\dfrac{y}{a}\right)^2 = 1$

B) $\dfrac{x^2}{a^2} + \dfrac{y^2}{a^2} = 1 \Rightarrow x^2 + y^2 = a^2$

We get the same equation as in 2c, with a equal to the radius. The dilation rules seem to work fine.

5) Here we move into new territory, but it should be easy travel for students. They are simply applying unequal vertical and horizontal dilations to a circle in order to create an ellipse.

A) $\left(\dfrac{x}{5}\right)^2 + \left(\dfrac{y}{3}\right)^2 = 1 \Rightarrow \dfrac{x^2}{25} + \dfrac{y^2}{9} = 1$

B) A horizontal dilation of scale factor 5, and a vertical dilation of scale factor 3

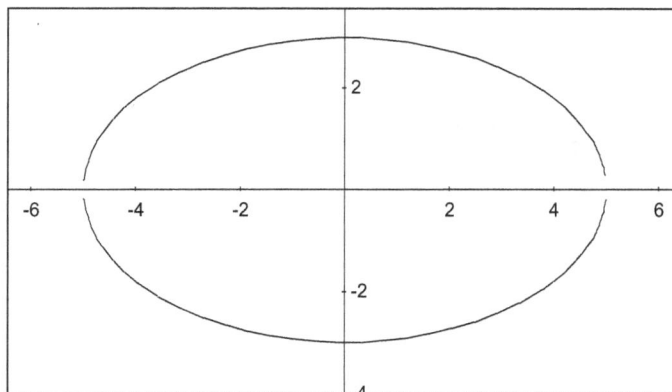

C)

D) $y = \pm\sqrt{9 - \dfrac{9x^2}{25}}$ Remind them that the +/-creates two functions: $y = \sqrt{9 - \dfrac{9x^2}{25}}$

and $y = -\sqrt{9 - \dfrac{9x^2}{25}}$ Each branch will produce half of the ellipse. Some students

may produce the equivalent form: $y = \pm 3\sqrt{1 - \dfrac{x^2}{25}}$.

6) Encourage students to sketch the graphs based on the characteristics, rather than on the computer this time.

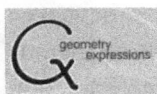

A) $\dfrac{(x+7)^2}{16} + \dfrac{(y-3)^2}{4} = 1$ Student graphs probably won't have the function equations.

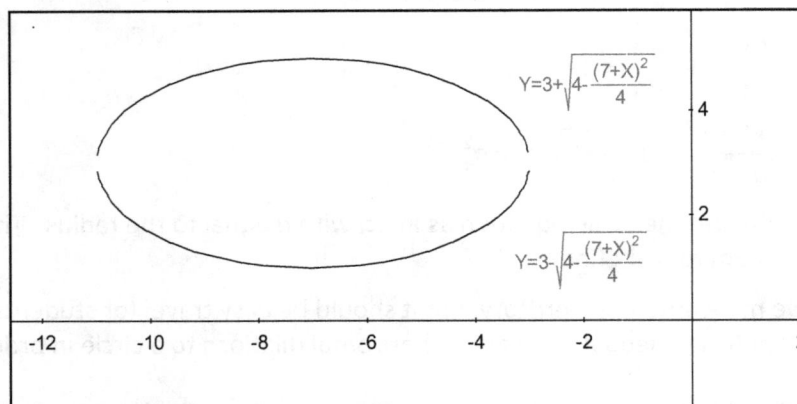

$Y = 3 + \sqrt{4 - \dfrac{(7+X)^2}{4}}$

$Y = 3 - \sqrt{4 - \dfrac{(7+X)^2}{4}}$

B) $\dfrac{(x-2)^2}{49} + \dfrac{(y+4)^2}{9} = 1$ Student graphs probably won't have the function equations.

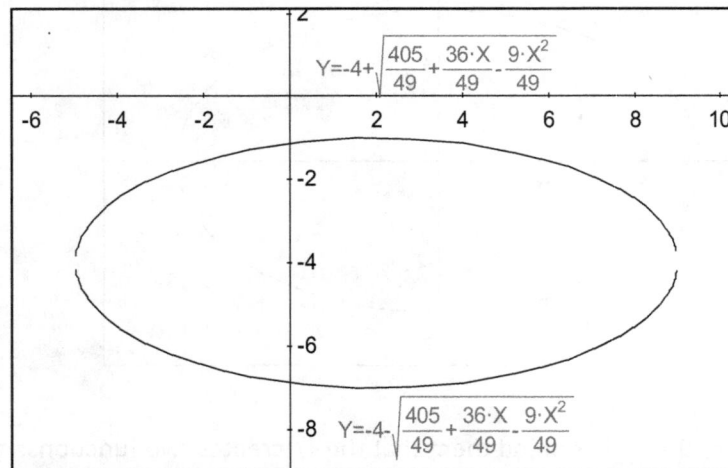

$Y = -4 + \sqrt{\dfrac{405}{49} + \dfrac{36 \cdot X}{49} - \dfrac{9 \cdot X^2}{49}}$

$Y = -4 - \sqrt{\dfrac{405}{49} + \dfrac{36 \cdot X}{49} - \dfrac{9 \cdot X^2}{49}}$

C) $\dfrac{(x-h)^2}{a^2} + \dfrac{(y-k)^2}{b^2} = 1$ or $\left(\dfrac{x-h}{a}\right)^2 + \left(\dfrac{y-k}{b}\right)^2 = 1$

Student Worksheets

Student worksheets follow.

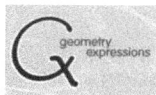

Function Transformations Ext. A: Circles and Ellipses

Today you are going to extend what you have learned about function transformations to graphs that don't represent functions. As you recall, in a function there is only one *y* value for any given *x* value. Graphically, this means that any vertical line will only cross the graph one time. Equations for functions can be written in the form *y* = . . . or *f(x)* = . . ., in which the right side of the equation is in terms of *x* alone. All the vertical and horizontal dilations and translations you have done so far have been done to functions. So what happens to other shapes, like circles?

1) First, we need an equation for a circle. Open a new GX file, and create a unit circle (radius = 1) with point A centered on the origin and point B on the circle. Draw in segment AB. Draw segment BC with C on the *x*-axis, and constrain BC to be perpendicular to the *x*-axis. You now have a right triangle with one vertex (B) on the unit circle.

A) What side of the triangle has a length equal to the *x*-coordinate of point B?

B) What side of the triangle has a length equal to the *y*-coordinate of point B?

C) What is the length of the hypotenuse?

D) Use the Pythagorean theorem to write an equation which relates *x*, *y*, and the length of the hypotenuse.

You can drag B around; the coordinates of all the possible points "B" describe the circle. These coordinates are described by the equation you wrote in part D.

2) What if the radius isn't 1?

A) Modify your diagram on GX from part 1 to have a radius of 2. Notice that this dilates the shape horizontally and vertically at the same time. More on this later. Use the Pythagorean theorem again to write a new equation which relates *x*, *y*, and the length of the hypotenuse.

B) Repeat the process for a circle of radius 3.

C)	Now consider the general case, a circle with radius r. Write an equation which relates x, y, and r. This is the general equation for a circle centered at the origin.

3)	The next logical question is what happens if the circle isn't centered on the origin. Change your radius back to 1, and then draw in a vector. Constrain your vector to represent a horizontal translation of 3 and a vertical translation of –2. Translate the circle and its center together.

A)	If the same translation rules hold for circles as for functions, what should the new equation be?

B)	Multiply out the expressions in parenthesis, and write your new equation here.

C)	Use known algebra properties to simplify the equation, and make the right side = zero. Write your new equation here.

D)	Highlight the image circle, and use **calculate real implicit equation** to check the equivalence. Do the same translation rules seem to apply to circles?

E)	Write the general equation for a circle with radius r, which has been translated h units right and k units up. Write in a form similar to part A, one that isn't multiplied out.

F)	Modify your diagram, replacing the vector coordinates with h for horizontal translation, and k for vertical translation. Use **calculate symbolic implicit equation** , and show algebraic steps to check the equivalence your equation from part E.

4)	We've already seen that changing the radius of a circle stretches the circle horizontally and vertically at the same time. How does this fit together with the dilations we've already done?

A)	Take the equation for a unit circle (from 1D), and replace x with $\dfrac{x}{a}$ and y with $\dfrac{y}{a}$.

B)   Simplify the equation and multiply both sides by a^2. What do you notice? Do the dilation rules seem to work the same way for circles?

5)   Now consider what happens if we dilate a circle vertically and horizontally by different scale factors. Obviously, the result will no longer be a circle.

   A)   Take the equation for the unit circle (from 1D) and replace x with $\frac{x}{5}$ and y with $\frac{y}{3}$. Write and simplify the equation below:

   B)   What transformations are represented by the changed equation in part A?

   C)   Sketch below what the equation from part A should look like if all the transformation rules hold true. This is called an ellipse. In general, an ellipse can be thought of as a "stretched out" circle, but it also has a number of special qualities that will become apparent when you study conic sections.

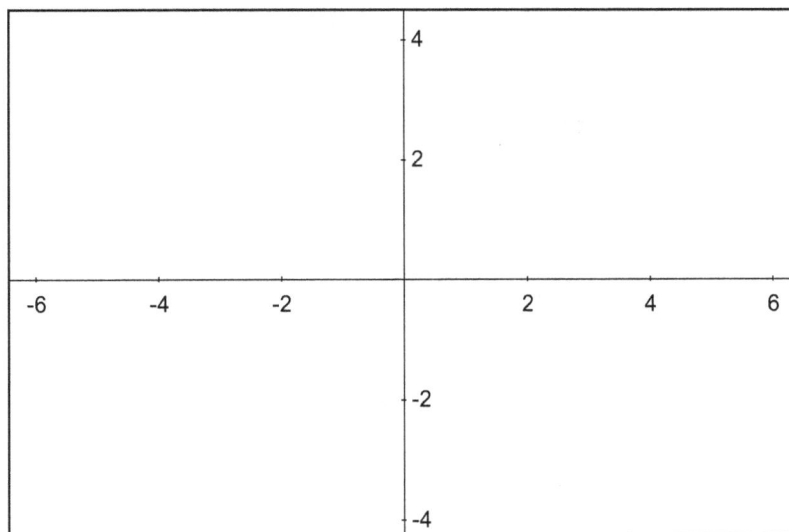

D) Check your work with GX. Solve the equation in part B for y. Remember that when you take the square root of both sides of an equation, you must insert +/-. This, in effect, creates two functions. Graph both of them using the function tool in GX. Write the equation(s) below:

6) Now put it all together:

A) Write an equation and sketch the graph of an ellipse which represents a horizontal dilation of scale factor 4, a vertical dilation of scale factor 2, and a translation up 3 and 7 units to the left.

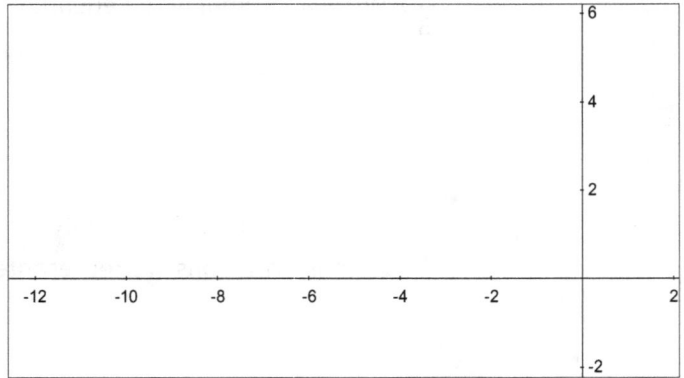

B) Write an equation and sketch the graph of an ellipse, which represents a horizontal dilation of scale factor 7, a vertical dilation of scale factor 3, and a translation down 4 units and 2 units to the right.

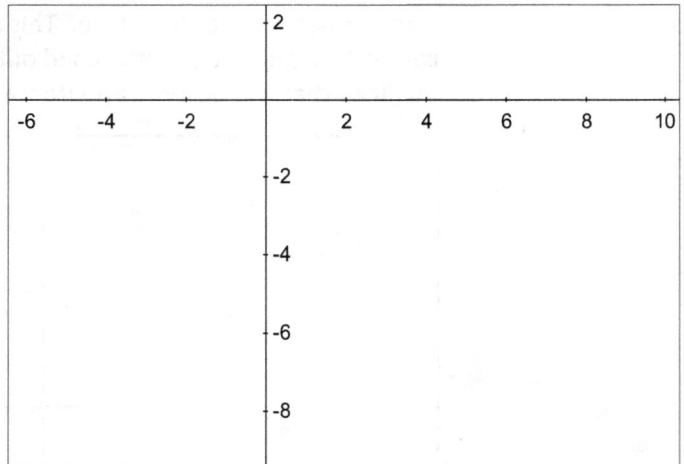

C) Write a general equation for an ellipse, which represents a horizontal dilation of scale factor a, a vertical dilation of scale factor b, and a vertical translation of k units and a horizontal translation of h units.

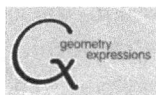

Extension B: Absolute Value

Learning Objectives

Math Objectives

- Students will apply transformations to absolute value functions, and analyze the effects on the piecewise slopes and on the vertex of the function.

Technology Objectives

- None

Math Prerequisites

- Successful completion of the core function transformations unit.

- Understand calculation of slope.

Technology Prerequisites

- Knowledge of Geometry Expressions as developed in the core function transformations unit.

Materials

- A computer with Geometry Expressions for each student or pair of students.

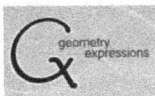

Overview for the Teacher

Function Transformations Ext. B: Absolute Value

This is a short exercise extending the transformation ideas to the absolute value function. It provides extra practice with the transformation concepts, and also gives students some experience with manipulating the absolute value function. The extension explores the relationships between horizontal and vertical dilations in three functions, and makes this more of a full-length lesson.

1) Students must draw in the infinite line in order to select one side of the graph to calculate slope on GX. A v-shape technically doesn't have a constant slope. Students can drag their points back and forth to determine future slopes. Also, you may have to clarify that at $x = 0$, the slope is undefined; a sharp corner can't be said to be pointed in a particular direction. This foreshadows differentiability questions in calculus.

A)

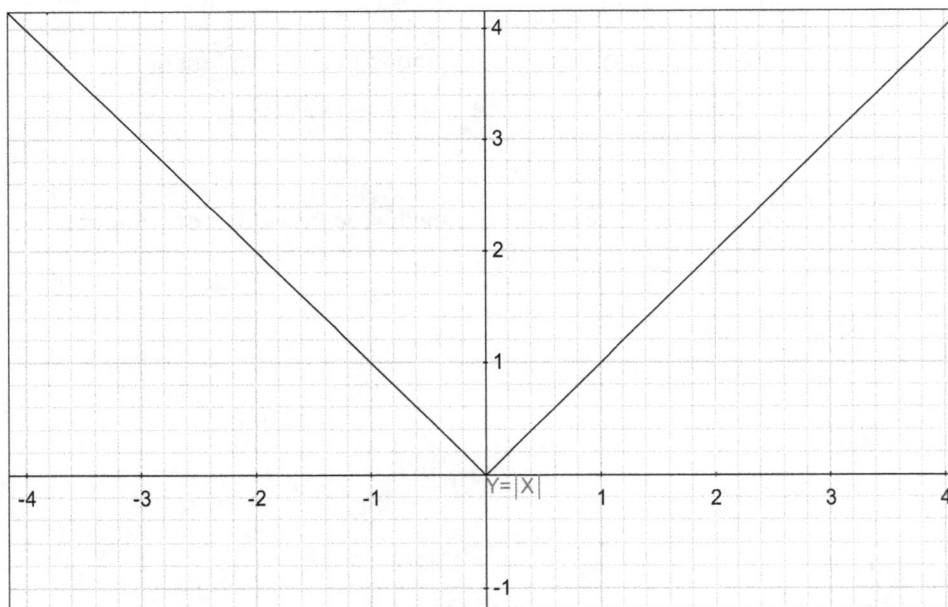

B)   1

C)   −1

D)   (0, 0)

Function Transformations - Extension B: Absolute Value
Algebra 2; Pre-calculus
Time required: 20 – 55 min.

2)

A) Horizontal translation of *h* units, vertical translation of *k* units.

B) Slope = 1 if *x* > 2 This is a key point: the value at which the domain of the piecewise function changes, the vertex of the graph. It shifts to the right.

C) Slope = -1 if *x* < 2

D) (2, 5) If students don't make the connection themselves, you may want to remind them of vertex form of quadratics.

3)

A) $y = 5 * |x|$ Notice that the true pattern of what we are doing is replacing *y* with $\frac{y}{5}$, which is why the 5 is outside the absolute value symbols. A scale factor inside would work equivalently if it were positive, but differently for negative numbers.

B) Slope = 5 if *x* > 0

C) Slope = -5 if *x* < 0

D) Vertex: (0, 0)

4)

A)

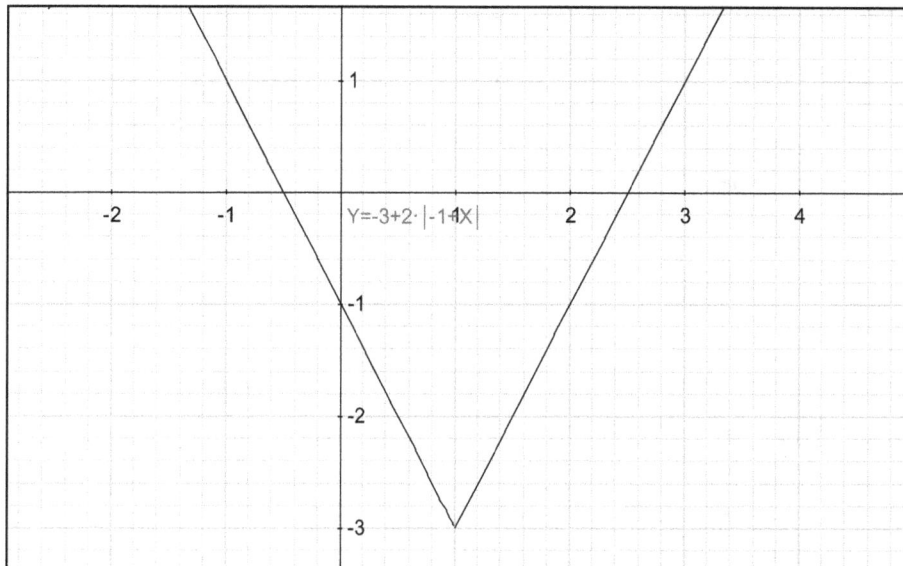

B) Slope = 2 if *x* > 1

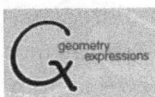

C) Slope = -2 if x < 1

D) Vertex: (1, -3)

5)

A) Slope = 3 if x > -2

B) Slope = -3 if x < -2

C) Vertex: (-2, 1)

D) $y = 3 * |x + 2| + 1$

EXTENSION:

A further possible extension would be to have students attempt horizontal dilations, which in this function often have equivalent vertical dilations. This could also be examined in function families $y = x^2$ and $y = \dfrac{1}{x}$. Students could explore models on GX and then work through the algebra to determine the relationship.

The key questions would be:

1) A vertical dilation of scale factor a corresponds to a horizontal dilation of _____?

Conversely, a horizontal dilation of b corresponds to a vertical dilation of _____?

2) Under what conditions does the rule in #1 hold true? Explain what happens when those conditions don't hold.

Answers:

$y = |x|$

1) A vertical dilation of scale factor a corresponds to a horizontal dilation of $\dfrac{1}{a}$.

Conversely, a horizontal dilation of b corresponds to a vertical dilation of $\dfrac{1}{b}$.

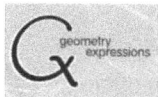

2)   This holds true if the values are positive numbers. A negative vertical dilation has no corresponding horizontal dilation. A negative horizontal dilation of b has a corresponding vertical dilation of $\dfrac{1}{|b|}$ or, since we know the sign of b, $\dfrac{1}{-b}$.

$y = x^2$

1)   A vertical dilation of scale factor a corresponds to a horizontal dilation of $\dfrac{1}{\sqrt{a}}$.

Conversely, a horizontal dilation of b corresponds to a vertical dilation of $\dfrac{1}{b^2}$.

2)   This holds true if the values are positive numbers. A negative vertical dilation has no corresponding horizontal dilation. A negative horizontal dilation of b has a corresponding vertical dilation of $\dfrac{1}{b^2}$, since squaring the number eliminates the effect of the negative.

$y = \dfrac{1}{x}$

1)   A vertical dilation of scale factor a corresponds to a horizontal dilation of a. Conversely, a horizontal dilation of b corresponds to a vertical dilation of b.

2)   This always holds true. Students can probably see graphically how stretching horizontally in this case is equivalent to stretching vertically. This is due to the symmetry across the line $y = x$. Put another way, the function is its own inverse. Students may also notice that GX automatically simplifies the equations to be equivalent. To work out the algebra, they will need to divide by a fraction.

Student Worksheets

Student worksheets follow

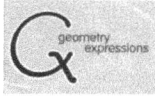

Name: _____

Date: _____

Function Transformations Ext. B: Absolute Value

To begin this activity, you must remember what absolute value $|x|$ means. In simple terms, absolute value is the distance from zero to a number on a number line. For example, $|5| = 5$ and $|-5| = 5$ as well. It is useful when a mathematical application has to have a positive value (like distances). A more formal definition for $y = |x|$ is that y = x for all $x \geq 0$ and y = -x for all $x \leq 0$. Remember that –x should be read "the opposite of x", and doesn't necessarily mean a negative number.

1)   Open a new GX file and graph $y = |x|$.

Type "abs(x)" for $|x|$ in GX.

[To determine slope: draw two points on one side of the function graph, then draw an infinite line through them. Highlight the line (not the function graph) and use **calculate real slope** .]

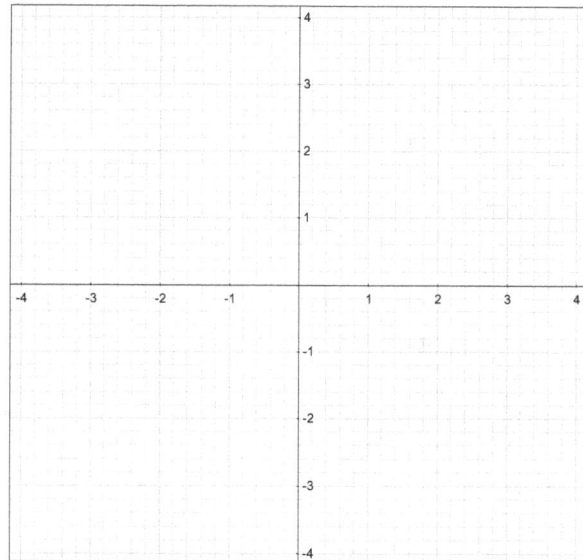

Note key characteristics below:
A) Sketch the graph.

B) slope = _____ if x > 0

C) slope = _____ if x < 0

D) Vertex is at point (,   )

2)   Now edit the graph to $y = |x - h| + k$ and test different h and k values.

A)   What effects do the h and k have on the graph?

Note key characteristics for h = 2 and k = 5:

B)   Slope = _____ if _____ (Be careful here; the pattern changes whenever the value inside the absolute value signs reaches zero.)

C)   Slope = _____ if _____

D)   Vertex is at point (,   )

3) Now we want to consider a vertical dilation.

A) Based on what you know about vertical dilations, write the equation for the absolute value function after it has been stretched vertically by a scale factor of 5.

Note the key characteristics:

B) Slope = _____ if _____

C) Slope = _____ if _____

D) Vertex is at (,)

4) Note the key characteristics and graph: $y = 2 * |x - 1| - 3$

A) Sketch the graph:

B) Slope = _____ if _____

C) Slope = _____ if _____

D) Vertex is at (,)

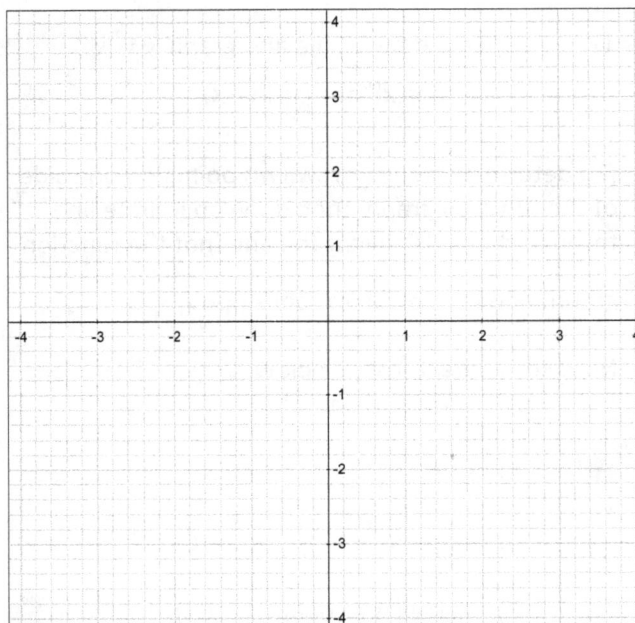

5) An absolute value function is shown in the graph to the right.

A) Slope = _____ if _____

B) Slope = _____ if _____

C) Vertex is at (,)

D) Write the equation for the function:

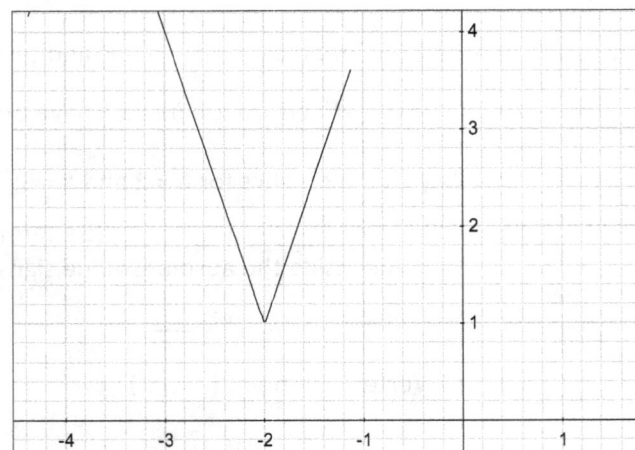

Extension C: Cosine and Tangent

Learning Objectives

Math Objectives

- Students will describe and simulate harmonic and circular motion using the cosine function.

- Students will apply horizontal shifts and translations to the tangent function.

Technology Objectives

- None.

Math Prerequisites

- Successful completion of the core function transformations unit.

- Understand calculation of slope.

Technology Prerequisites

- Knowledge of Geometry Expressions as developed in the core function transformations unit.

Materials

- A computer with Geometry Expressions for each student or pair of students.

Overview for the Teacher

Function Transformations Ext. C: Cosine and Tangent

This set of exercises gives yet another opportunity for students to review and practice transformations of functions. It also gives them the opportunity to become more comfortable with the trigonometry function graphs, which will serve them well as they move into higher mathematics.

1)

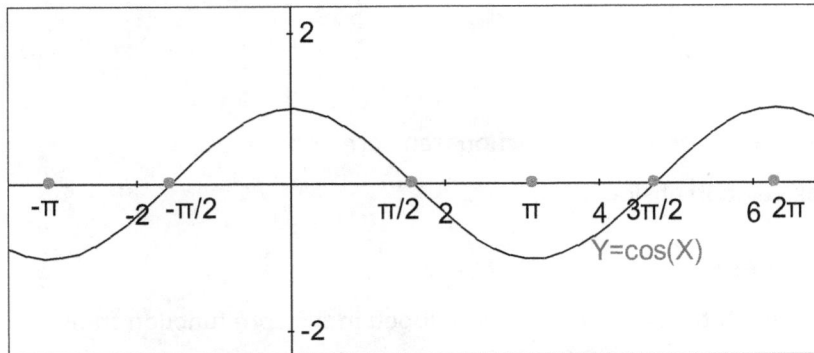

2)

A) a: Vertical dilation of scale factor a; also amplitude of the function.

B) P: Period of the function

C) h: Horizontal translation; also called phase shift

D) k: Vertical translation

3)   This should be a relatively easy repeat of the work they did with sine.

A) $y = 20*\cos\left(\dfrac{\pi}{11}*t\right) + 25$ Note that the horizontal translation is zero. It may be beneficial to point out to students that this equation is simpler than the sine equivalent because it lacks the horizontal translation.

B)   First diagram is the computer screen. Check that their animations work; points C and D should be at the same height at all times.

Function Transformations Extension C - Cosine and Tangent
Algebra 2; Pre-Calculus
Time required 50 – 75 min.

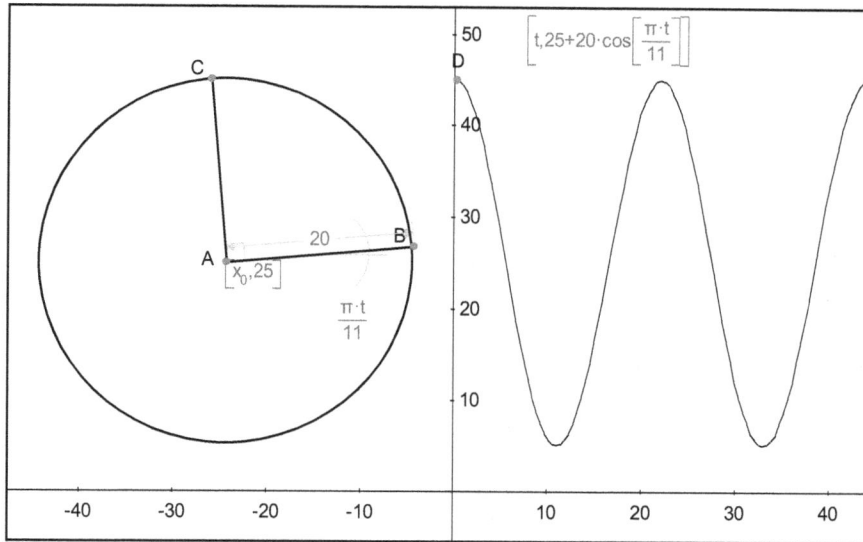

C)

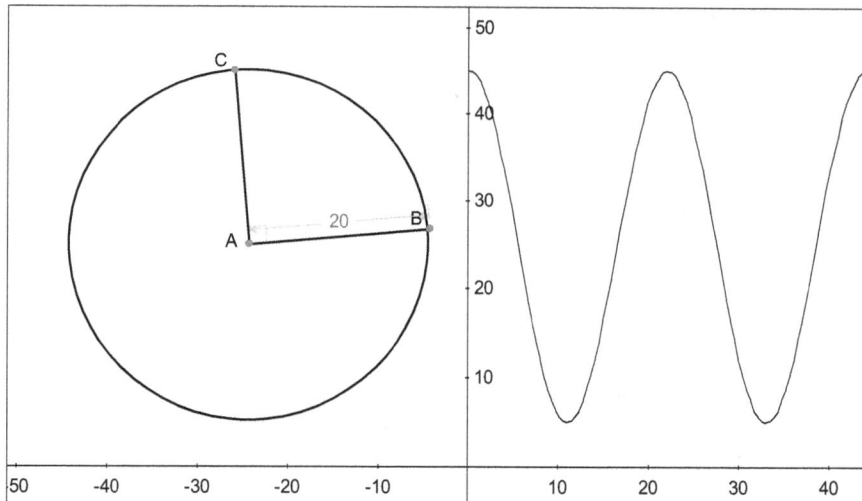

4) Students may recognize this problem from the work they did with sine. That fact will reinforce the relationship between the two functions. Students must do some interpretation of the given information to turn it into the parameters they need. If the weight goes from its lowest to its highest point in 1.5 seconds, the whole period must be 3 seconds. The vertical dilation is 0.7 – the difference between the equilibrium height and the lowest point.

A)

- $a = 0.7$

- h = 1.5 or h = −1.5; Since the starting point is exactly halfway through the cycle from cosine's natural starting point, the translation works the same left or right.

- k = -1.2

- P = 3

B) $y = 0.7 * \cos\left(\frac{2\pi}{3}(x - 1.5)\right) - 1.2$

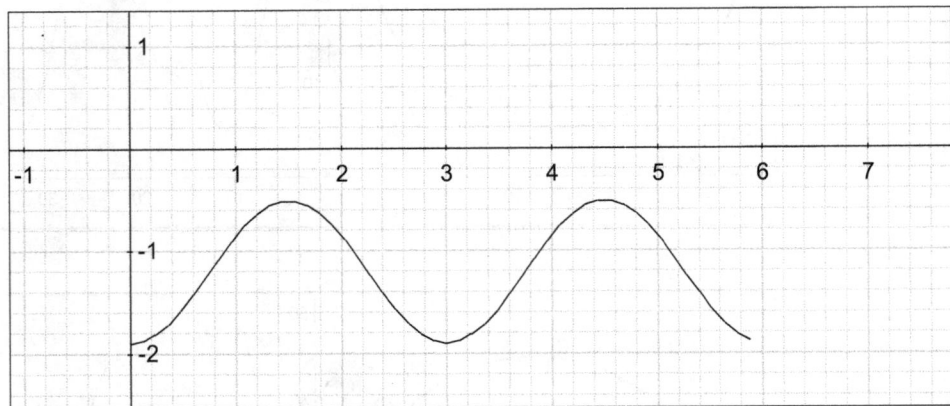

5) A) The slope values get very large

B) The slopes go to extreme negative values.

C) infinity

D) Tangent repeats every π radians, or 180 degrees. If students have trouble seeing this, have them construct a rotation of line segment AB, with the angle of rotation being π radians. They can then calculate the slope of segment AB', and drag point B around. They should quickly see that all different angles give the same slope for AB' as they do for AB.

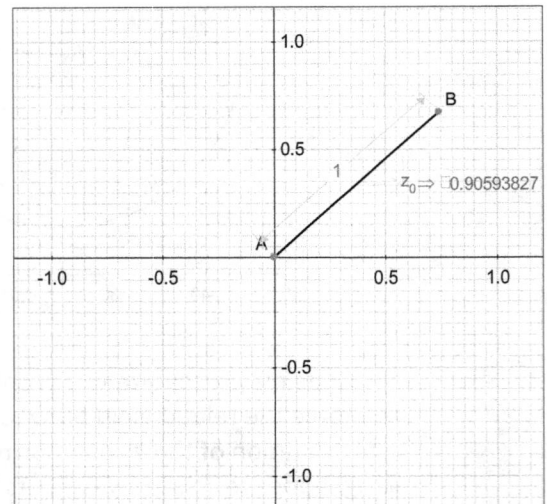

E)

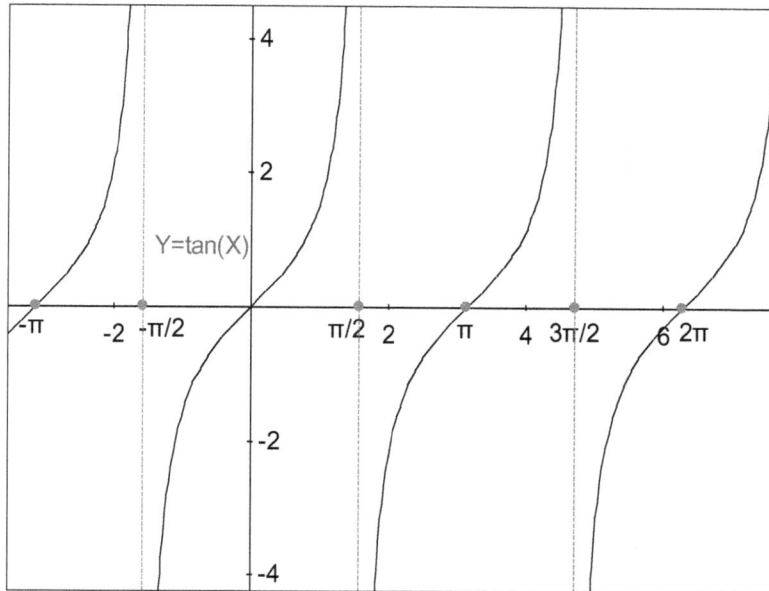

6)   A) horizontal translation of $\dfrac{\pi}{2}$

B)

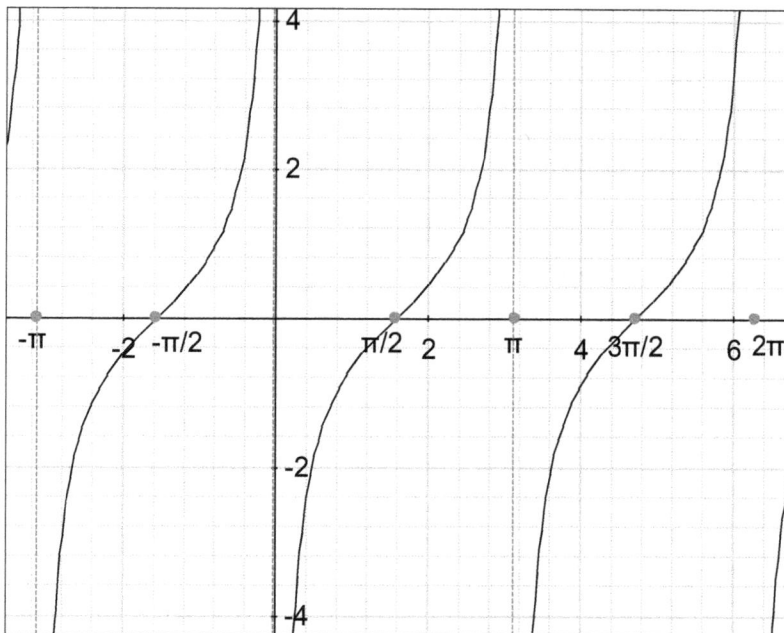

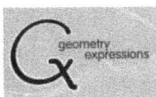

7)   A) A horizontal dilation of scale factor 2; double the period.

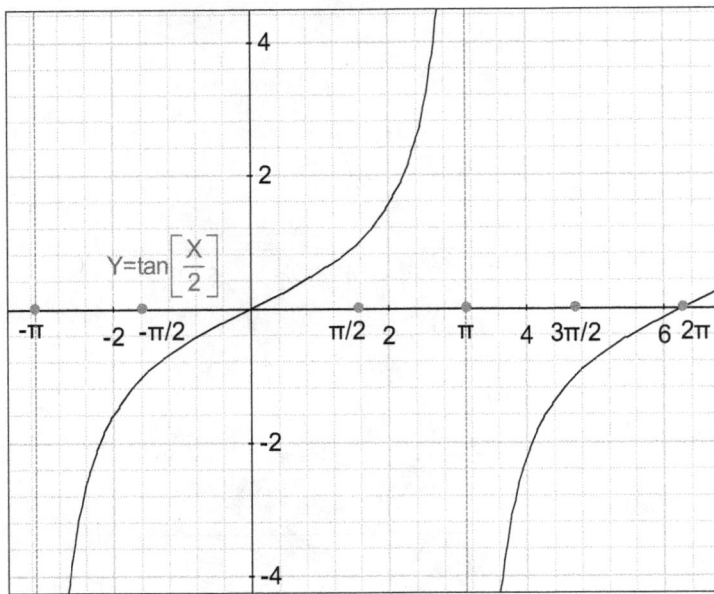

    B)

$Y = \tan\left[\dfrac{X}{2}\right]$

Student Worksheets

Student worksheets follow.

Function Transformations Ext. C: Cosine and Tangent

Quick Review of Trig Functions:

On a right triangle, $\sin(\theta) = \dfrac{Opposite}{Hypotenuse}$, which also corresponds to the y-coordinate of point B on the unit circle to the right. Similarly, $\cos(\theta) = \dfrac{Adjacent}{Hypotenuse}$ and the x coordinate on the circle. $Tan(\theta) = \dfrac{Opposite}{Adjacent} = \dfrac{y}{x}$. This corresponds to the slope of line segment AB, which we'll consider more later.

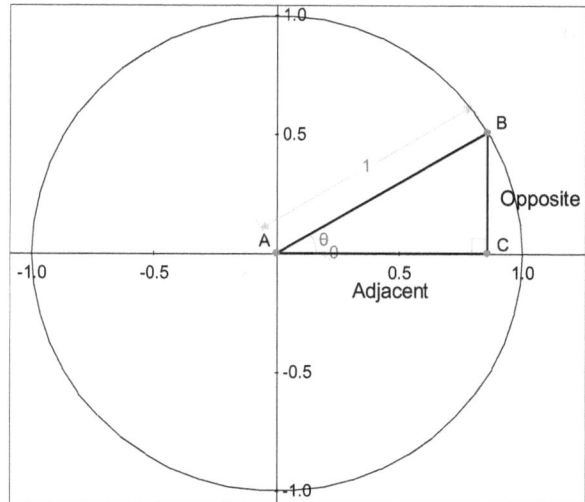

In previous simulations, you used the sine function, which worked out nicely for modeling heights. You further noticed that the cosine function is simply a horizontal shift of $-\dfrac{\pi}{2}$ from the sine function. This is a result of the fact that the x- and y- axes are a $90°$ or $\dfrac{\pi}{2}$ radians rotation of each other. This means that every model we did with the sine function could also be done with a cosine function.

1)   Sketch a graph of $y = \cos(\theta)$. This can be done from memory or by using GX.

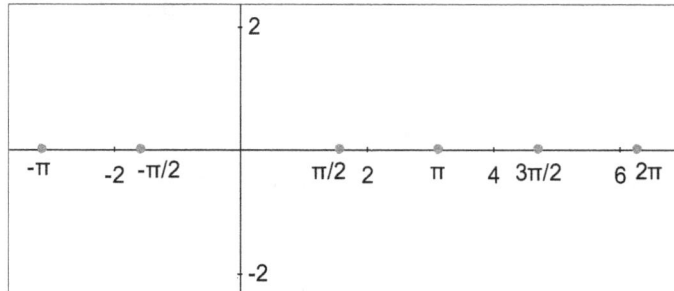

2)   The generic form of a cosine function is given by $y = a * \cos\left(\dfrac{2\pi}{P} * (x - h)\right) + k$. Give the significance of each of the parameters.

A)   *a*:

B)   *P*:

C)   *h*:

D)   *k*:

3)   A small Ferris wheel is centered 25 feet above the ground, and has a radius of 20 feet. It takes 22 seconds to complete one full counter-clockwise revolution. You are tracking the distance from the ground of a child who was at the peak of the ride when it started moving.

A)   Write an equation for the child's height as a function of time using cosine.

B)   Create a simulation of this motion in GX, and create a simultaneous construction of the function graph. This will be done almost exactly like the sine simulations:

- Construct a circle with the appropriate radius and center height.

- Draw a line segment from the center to the circle.

- Constrain the direction of your line segment to be the expression you have in parenthesis after cosine.

- The difference is you need to construct a second line segment from the center to the edge of your circle, and constrain it to be perpendicular with the first. This gives you the $90°$ rotation, which converts from sine to cosine. This second point represents the position of the child you are tracking.

- Plot a point, and constrain its coordinates to (*t*, your equation from part A)

- Run a simulation for two periods, and construct the locus of the point.

Teacher Initials: _____

C)   Copy the graph and simulation diagram:

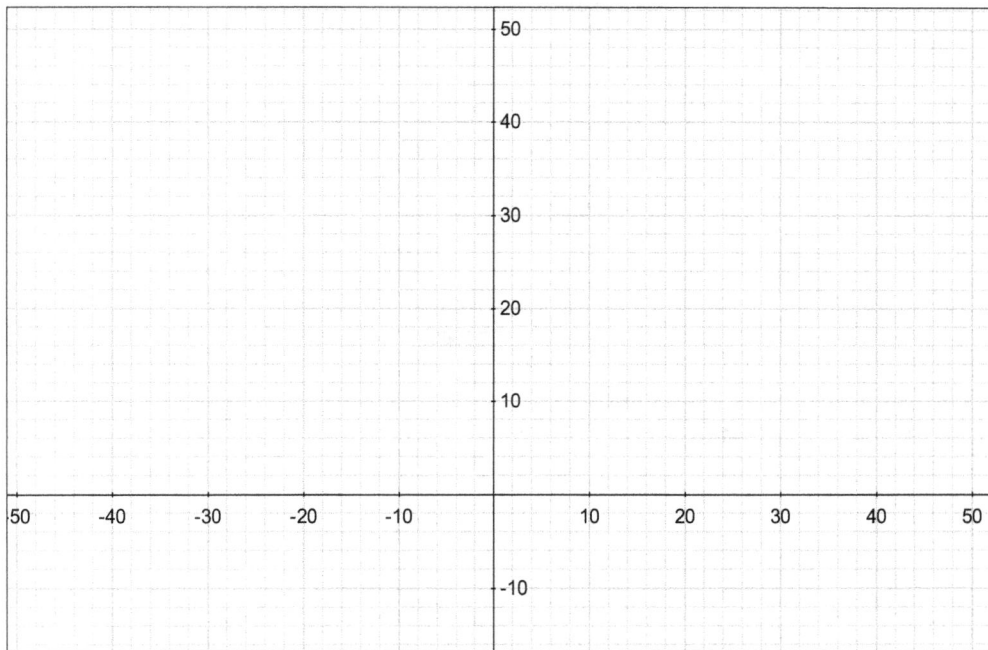

4)   A weight is hanging on a spring below a board. It is stretched and released, creating a harmonic motion pattern as the weight oscillates up and down. At equilibrium, the weight rests 1.2 meters below the board. It is released at time zero from its maximum stretch, which puts the weight 1.9 meters below the board. It reaches its maximum height of 0.5 meters below the board after 1.5 seconds.

    A)   Identify the values of the four parameters of a cosine curve for this example.

        • $a =$

        • $h =$

        • $k =$

        • $P =$

    B)   Write the equation for the function, using cosine.

    C)   Now open a new GX file and create a simulation of the weight. This will simply look like a point oscillating up and down. Set the animation to run the simulation through two complete cycles.

D)   Now create the function simultaneously with the simulation, like you did in problem #3. Copy your function graph below.

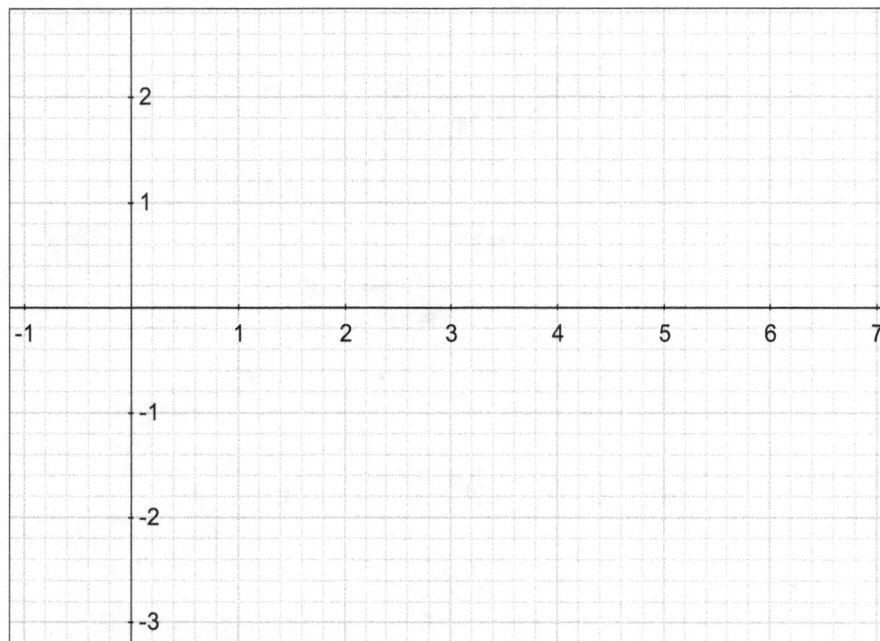

Tangent Function:

The tangent function, as we noted earlier, equals $\frac{Opposite}{Adjacent}$, and corresponds to the slope of line segment AB of our earlier diagram of the unit circle.

5)   Construct a simplified model of the unit circle: Draw line segment AB with A constrained to the origin, and constrain its length to be one. Use **calculate real slope** to determine the slope at any given angle, and drag point B around.

   A)   What happens to the slope (tangent value) as you drag B in the first quadrant, very close to the y axis?

   B)   What happens to the slope (tangent value) as you drag B slightly past the y axis?

   C)   Temporarily constrain the angle AB makes with the x-axis to be $\frac{\pi}{2}$, then delete the constraint. What was the slope (tangent value) at that angle? This suggests the need for an asymptote in the function graph.

D)	The sine and cosine patterns repeat every 2π. By dragging point B and experimenting, determine how often the tangent (slopes) will repeat. This is the period of the tangent function.

E)	Now sketch a graph of the tangent function. You may use GX to help. Draw in the asymptotes as dotted lines.

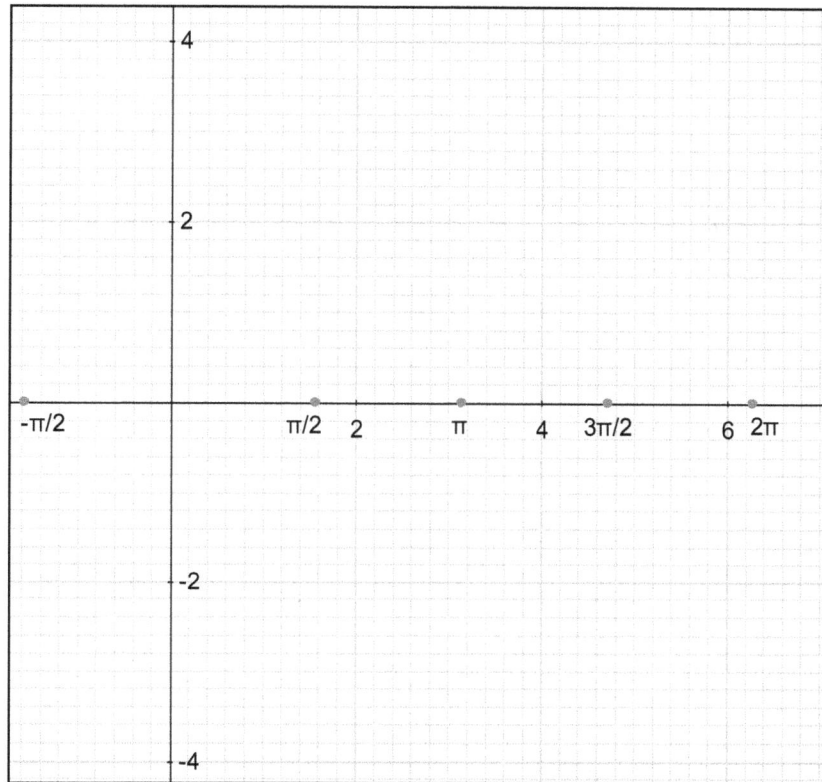

6)	Consider the function $y = \tan\left(x - \dfrac{\pi}{2}\right)$

A)	What effect will this change in the equation have on the tangent graph?

B)   Sketch the graph of the new function.

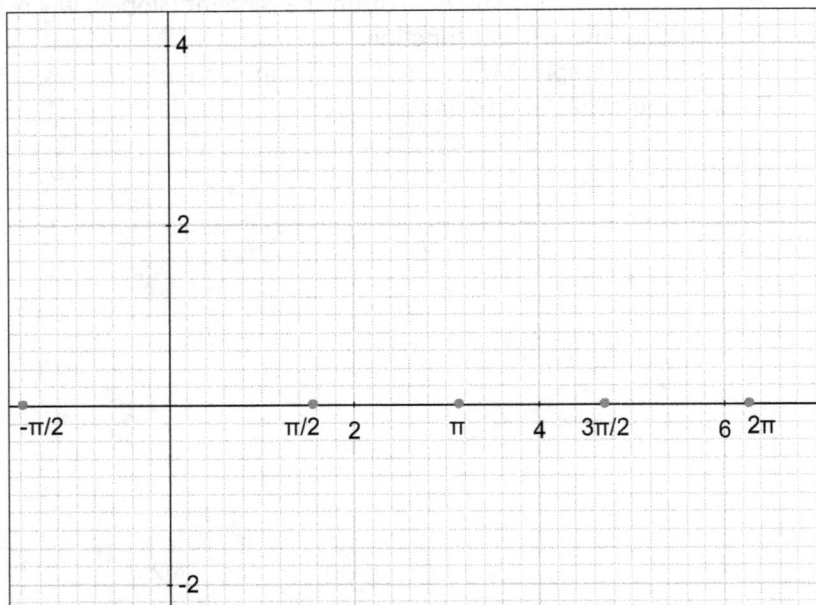

7)   Finally, consider the function $y = \tan\left(\dfrac{x}{2}\right)$.

C)   What effect will this change in the equation have on the tangent graph?

D)   Sketch the graph of the new function.

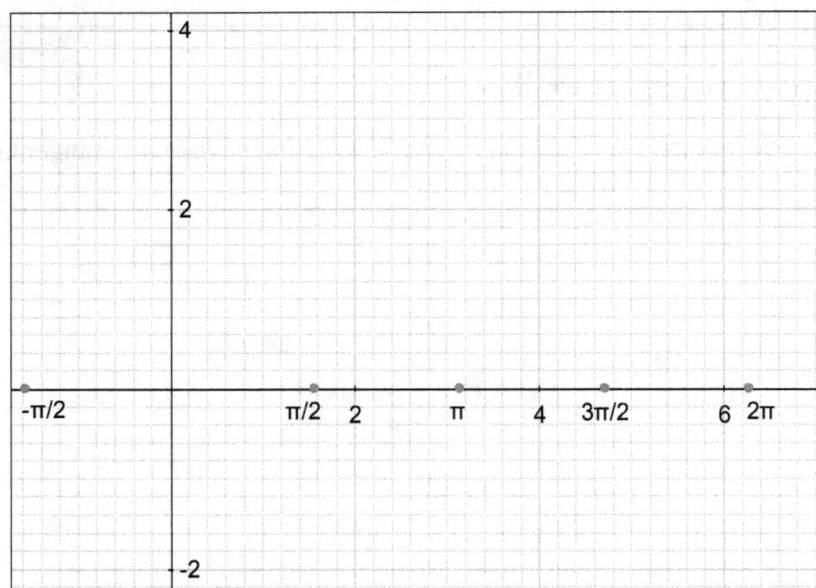

Extension D: Vertical Asymptotes

Learning Objectives

Math Objectives

- Students will understand that vertical asymptotes occur when the denominator of a function approaches zero.

- Students will be able to determine vertical asymptotes from the equations of rational functions in factored form.

- Students will gain an introductory familiarity with the secant, cosecant, and cotangent functions.

Technology Objectives

- None.

Math Prerequisites

- Successful completion of the core function transformations unit.

- Knowledge that $\dfrac{n}{0}$ and $\dfrac{0}{0}$ are both undefined.

- Understand and be able to use the definition of reciprocal.

Technology Prerequisites

- Knowledge of Geometry Expressions as developed in the core function transformations unit.

Materials

- A computer with Geometry Expressions for each student or pair of students.

- Colored pencils are optional, but helpful.

Overview for the Teacher

Students explore vertical asymptotes and why they happen in terms of the denominator of a fraction. In the process, they gain an introduction to rational functions as well as the secant, cosecant, and cotangent functions.

Students may want to color-code their graphs. It will improve the clarity of the ideas and the drawings, since they will be putting more than one function on a single set of axes.

1) This should be fairly easy for students. If they don't get a vertical line, have them make sure their computer is set to radian mode: Edit/Preferences/Math/Math/Angle Mode/Radians. You may have to remind them what reciprocal means.

   A) $x = 0$

   B) Answers may vary; this is just conjecture at this point. The asymptote goes through where the function equals zero. The asymptote also goes through the vertex, but this is true by coincidence, as they will see in exercise 3 & 4).

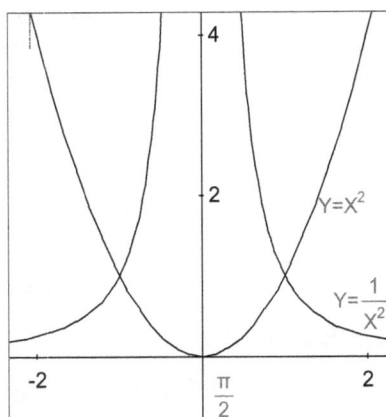

2)

   A) $x = 2$

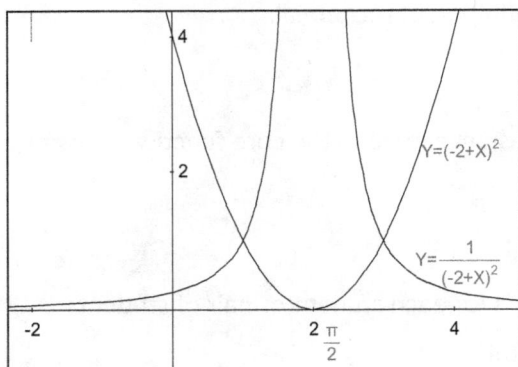

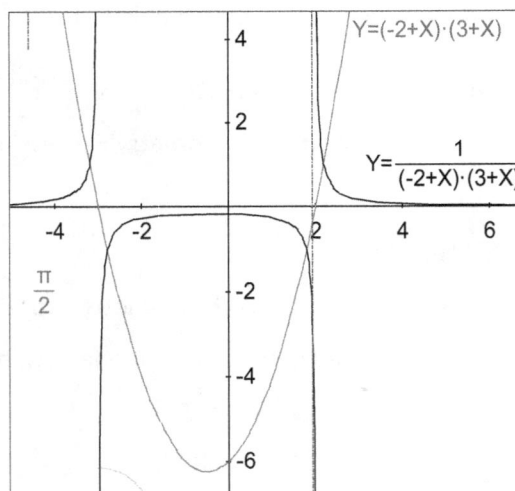

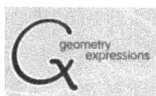

3)　You may need to remind students to type in * for multiplication in Geometry Expressions. One way to help students conceptualize this change is to note that the function in part 2 can be thought of as having two asymptotes which coincided. In this function, each of the two asymptotes was translated a different distance and direction.

A)　$x = 2; x = -3$

B)　Answers may still vary. Ideally students notice that the asymptotes are the zeros of the function reciprocal to $g(x)$.

4)　This problem is included to help students avoid tying the idea of vertical asymptotes to properties of parabolas, such as line of symmetry. Since the reciprocal of h is not a parabola, this should help remove any such misconceptions.

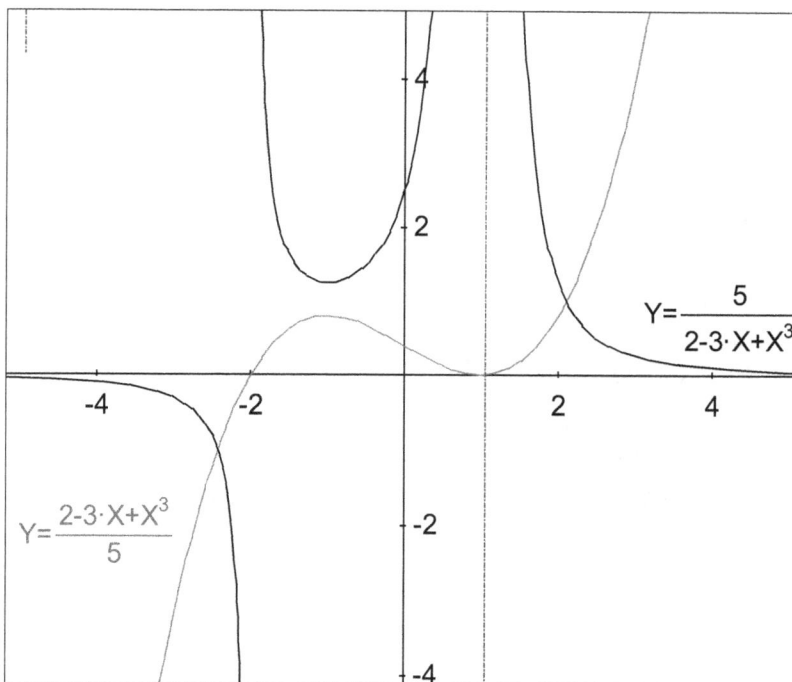

$$Y = \frac{5}{2 - 3 \cdot X + X^3}$$

$$Y = \frac{2 - 3 \cdot X + X^3}{5}$$

A)　$x = -2, x = 1$

B)　The asymptotes of $h(x)$ are located where the reciprocal of $h(x)$ equals zero.

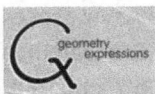

5)   This example yields non-integer answers.

A)

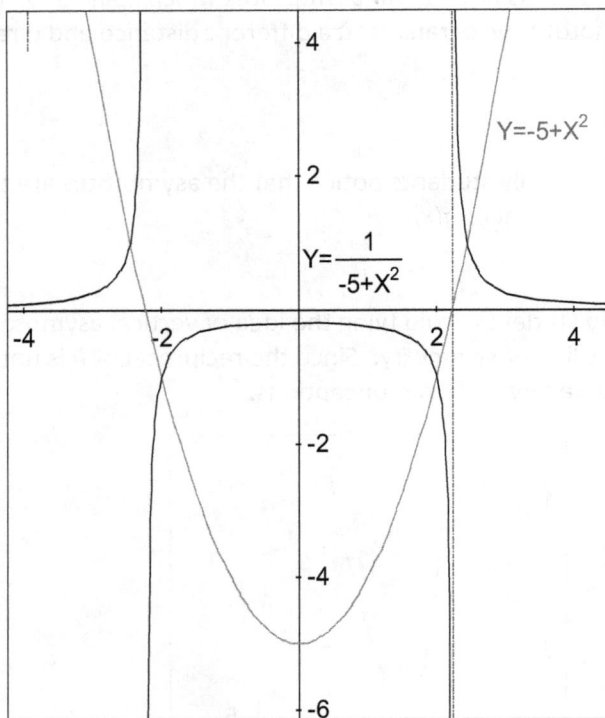

B) $x = \sqrt{5}$ and $x = -\sqrt{5}$ Students will probably have decimal approximations based on their GX work. This will be rectified in problem 5.

C)   Student conjectures should be correct at this point. The asymptotes correspond to the zeros of the function that is reciprocal to *j(x)*. This will develop more specifically into just looking at the denominator in problem 6.

6)

A)

x	10	1	0.1	0.01	0.001
$\dfrac{1}{x}$	0.1	0	10	100	1000

B)   The value of *y* gets extremely large.

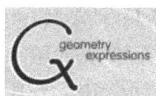

C)

x	-10	-1	-0.1	-0.01	-0.001
$\dfrac{1}{x}$	-0.1	-1	-10	-100	-1000

D) The value of y gets extremely small / the absolute value of y gets extremely large.

E) If x = 0, y is undefined.

At this point, you may want to raise the question with students of how to get exact answers for the asymptotes in problem 5. Someone will probably come up with solving the equation $0 = x^2 - 5$.

7) This exercise is a very basic introduction to the idea of a hole in a function's domain. It is included here simply to keep students from over-generalizing. They need to be aware that, while the pattern for zeros of denominators described above is generally true, it has exceptions.

A) Students may expect to see a vertical asymptote at x = -3. If some have figured out ahead of time that this won't happen, simply refer them to part B.

B) The (x + 3) terms in the numerator and denominator of the reciprocal of k(x) cancel out. This eliminates the vertical asymptote, and makes the reciprocal of k(x) behave as a straight line. GX simplifies this automatically, and ignores the issue of a hole.

We now move to analyzing trigonometric functions in terms of asymptotes. This serves both to reinforce the principles of vertical asymptotes, and to introduce secant, cosecant, and cotangent functions.

8) If the class did Extension C, they already know where the vertical asymptotes of the tangent function are. This provides a review, and an alternate way of understanding them. This new way aligns closely with principles of vertical asymptotes in other functions.

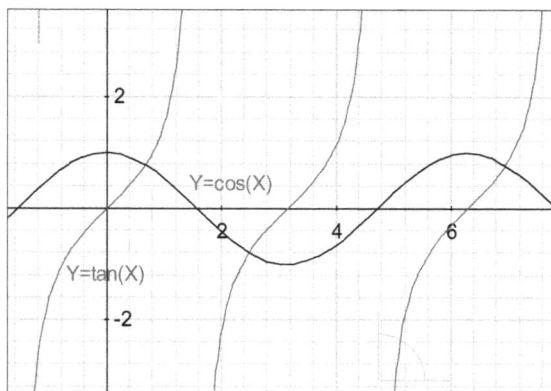

A)   Wherever the cosine function equals zero; specifically at $-\dfrac{\pi}{2}, \dfrac{\pi}{2}, \dfrac{3\pi}{2}, \dfrac{5\pi}{2}$, etc.

9)

A)   Wherever the cosine function equals zero; specifically at $-\dfrac{\pi}{2}, \dfrac{\pi}{2}, \dfrac{3\pi}{2}, \dfrac{5\pi}{2}$, etc.

Students should notice that these are the same as for the tangent function.

B)

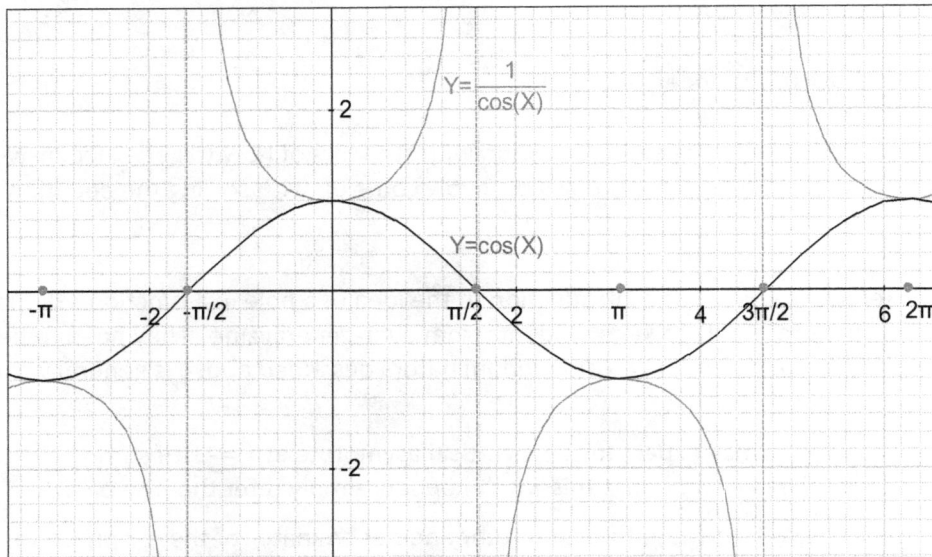

C)   Secant approaches 1

D)   Secant approaches –1

10)

A)   Wherever the sine function equals zero; specifically at $-\pi$, 0, π, 2π, etc.

B)

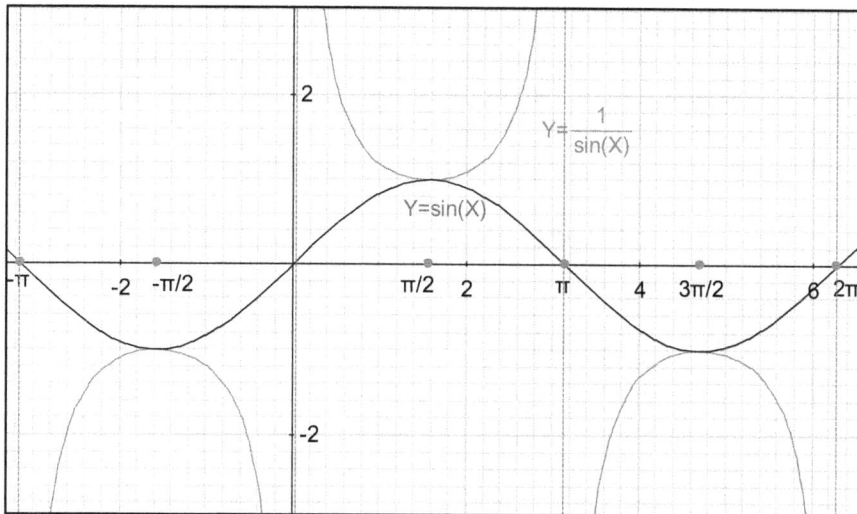

11)

A) Both functions have values of zero at the same points.

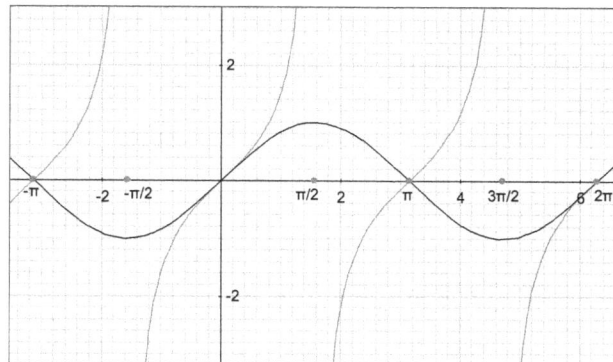

B) At $-\pi$, 0, π, 2π, etc. Students should notice that these are the same as for the cosecant function.

C)

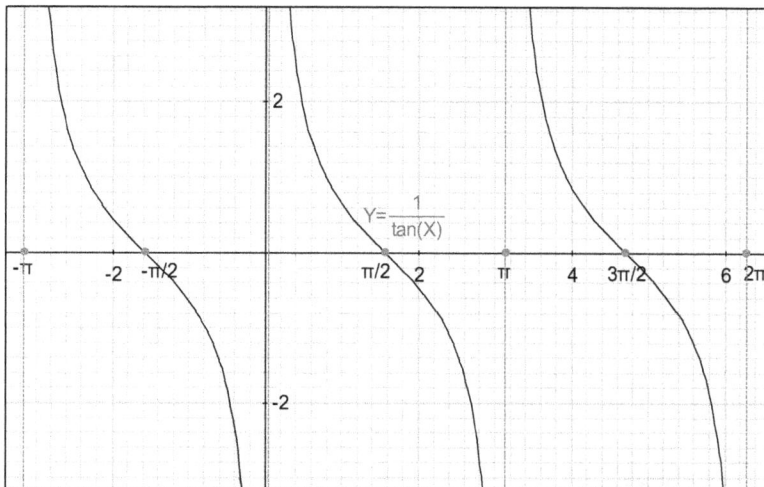

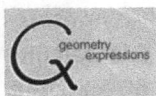

12)   "In general, you can expect to find a vertical asymptote in a function graph when the denominator of a function is zero." You may want to encourage students to articulate the exception to this rule – if it is a value of x that makes both the numerator and the denominator zero, it creates a hole. However, students shouldn't be expected to have the idea of the hole mastered from the minimal introduction in this lesson alone.

This lesson could lead into an in-depth look at rational functions, or simply serve as an introduction/familiarization to the concept. It doesn't cover the skills of factoring, the concept of end behavior, or the zeros of a rational function, and gives only a cursory look at holes. All of these ideas could be included in a progression from this starting point.

Student Worksheets

Student worksheets follow.

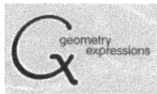

Name: _____

Date: _____

Function Transformations Ext. D: Vertical Asymptotes

You have looked at vertical asymptotes in the family of functions whose parent is $y = \frac{1}{x}$. Today you are going to look at some other functions with vertical asymptotes.

1)   Consider the function $y = \frac{1}{x^2}$. Open a new GX file and use the function tool to graph it.

       Also draw an infinite line, and constrain its direction to $\frac{\pi}{2}$, creating a vertical line.

       Make sure your computer is set to radian mode. Highlight it, right click, and select properties to change it to a dashed or dotted line. This will serve to represent your adjustable vertical asymptote for the rest of the exercises. Drag it to the place that appears to be a vertical asymptote in your graph.

       A)   What is the approximate equation for your vertical asymptote?

       B)   Now graph the reciprocal function: $y = x^2$. What do you notice about the asymptote compared to this function?

2)   Consider the function $f(x) = \frac{1}{(x-2)^2}$.

       A)   Based on what you know about translations, where should the vertical asymptote be?

       B)   Graph the equation, and the reciprocal of f, $y = (x-2)^2$, in GX. Copy the graphs of both functions, with the asymptote as a dotted line.

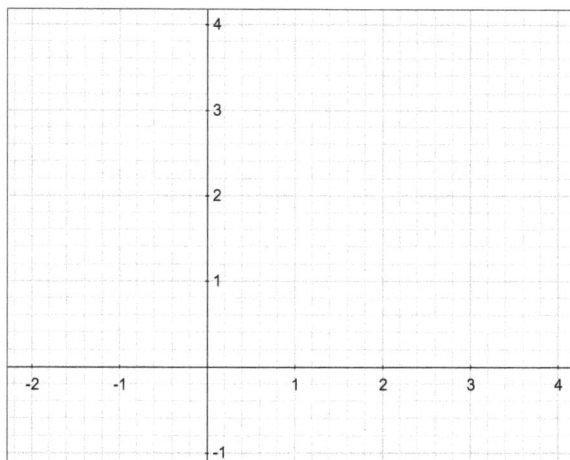

3) Now consider what happens when you modify the equation from $f(x) = \dfrac{1}{(x-2)(x-2)}$ to

create $g(x) = \dfrac{1}{(x-2)(x+3)}$. Graph $g(x)$ with the reciprocal of $g(x)$ on your computer.

A) You notice that you now have more than one vertical asymptote. Give the equations for both the vertical asymptotes.

B) What do you notice about the asymptotes compared to the reciprocal of $g(x)$?

4) Consider the function
$h(x) = \dfrac{1}{x^3 - 3x + 2}$.

A) Graph $h(x)$ and the reciprocal of $h(x)$, together with the vertical asymptotes.

B) Write the approximate equations for the vertical asymptotes.

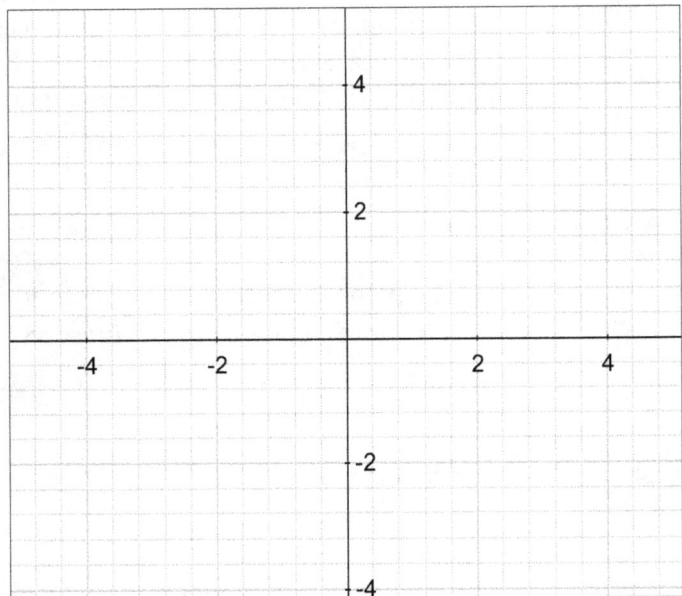

C) What do you notice about the asymptotes compared to the reciprocal of $h(x)$?

5) Consider the function $j(x) = \dfrac{1}{x^2 - 5}$.

    A)   Graph *j(x)* and the reciprocal of *j(x)*, together with vertical asymptotes.

    B)   Write the approximate equations for the vertical asymptotes.

    C)   What do you notice about the asymptotes compared to the reciprocal of *j(x)*?

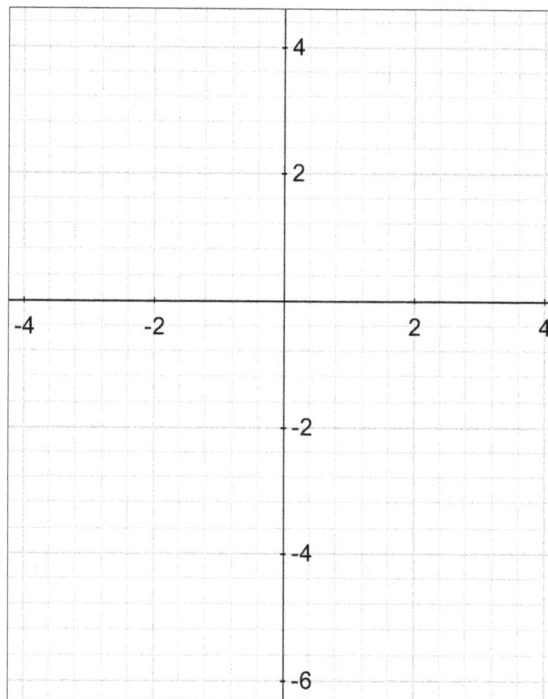

Why does this work?

6) We will consider the simplest function with a vertical asymptote: $y = \dfrac{1}{x}$.

    A)   Complete the table:

X	10	1	0.1	0.01	0.001
$\dfrac{1}{x}$					

    B)   What happens to *y* as *x* is positive, and gets closer and closer to zero?

    C)   Complete the table:

X	-10	-1	-0.1	-0.01	-0.001
$\dfrac{1}{x}$					

D)   What happens to *y* as *x* is negative, and gets closer and closer to zero?

E)   What happens when *x* = 0?

In general, if the value of the denominator approaches zero, the absolute value of the overall quotient becomes infinitely great. When the value of the denominator equals zero, the function is undefined, thus creating the asymptote.

7)   Consider the function $k(x) = \dfrac{x+3}{(x-2)(x+3)}$. You'll need four sets of parenthesis to type this in correctly: (x+3)/((x-2)*(x+3)). Graph *k(x)* and *y* = (x-2)*(x+3) on GX.

A)   What surprises you about this graph? I.e. where might you have expected to see an asymptote?

B)   Now graph the reciprocal of *k(x)*. Why is there a difference between the zeros of the reciprocal function, and the zeros of the function *y* = (x-2)*(x+3)? A look at the way GX simplifies the equation may help.

Both *k(x)* and its reciprocal behave as if the (*x* + 3) term didn't exist. The only exception to this pattern is when *x* = -3, where both functions are undefined, since one can't divide by zero. This is called a "hole" in the function.

8)   Now we're going to look at some trigonometric functions. Remember that $\tan(x) = \dfrac{\sin(x)}{\cos(x)}$. Open a new GX file and graph *y* = cos(x). Make sure your computer is in radian mode.

A)   Where would you expect to see a vertical asymptote in the tangent function? (More than one answer.) Construct your generic vertical asymptote as in the last file, and move it to one of the appropriate places.

B)   Now graph *y* = tan(x) in the same file. Does your expectation seem to be true? Move your asymptote to the various zeros of the cosine function to check.

There are three trigonometric functions that are defined as the reciprocals of sine, cosine and tangent. They are

- secant, which is $\sec(x) = \dfrac{hypotenuse}{adjacent} = \dfrac{1}{\cos(x)}$

- cosecant, which is $\csc(x) = \dfrac{hypotenuse}{opposite} = \dfrac{1}{\sin(x)}$

- cotangent, which is $\cot(x) = \dfrac{adjacent}{opposite} = \dfrac{\cos(x)}{\sin(x)} = \dfrac{1}{\tan(x)}$

9)   We'll start with the secant function. Delete tangent from your GX drawing, and consider the cosine function again.

    A)   Where would you expect to see vertical asymptotes in the secant function? (more than one answer.) Move your asymptote to one of the appropriate places

    B)   Graph $y = \sec(x)$ in the same file. You have to type this in as 1/(cos(x)). Copy the graph of both functions here, including asymptotes.

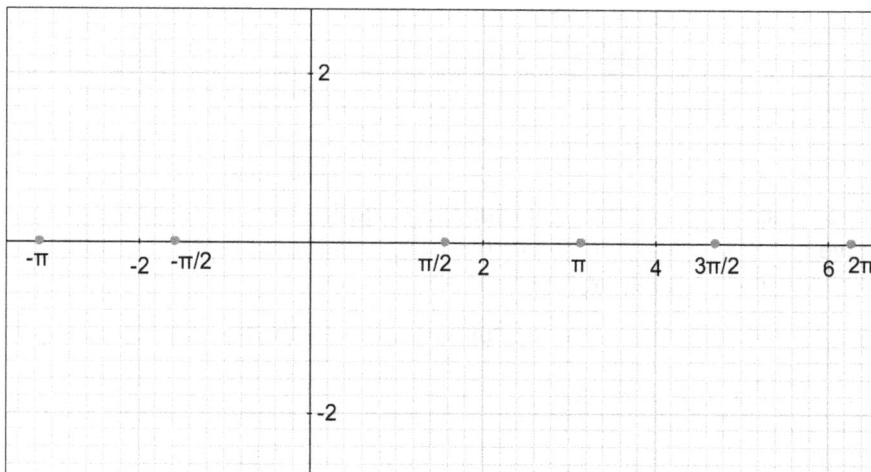

    C)   What happens to the secant function as the cosine function approaches a value of 1 or –1 (i.e. at $x = 0$, $x = \pi$, and $x = 2\pi$)?

10) Now look at the cosecant function. Delete secant, and change the cosine function to the sine function on your drawing.

A) Where would you expect to see vertical asymptotes in the cosecant function? (More than one answer.) Move your asymptote to one of the appropriate places.

B) Graph $y = \csc(x)$ in the same file. You have to type this in as $1/(\sin(x))$. Copy the graph of both functions here, including asymptotes.

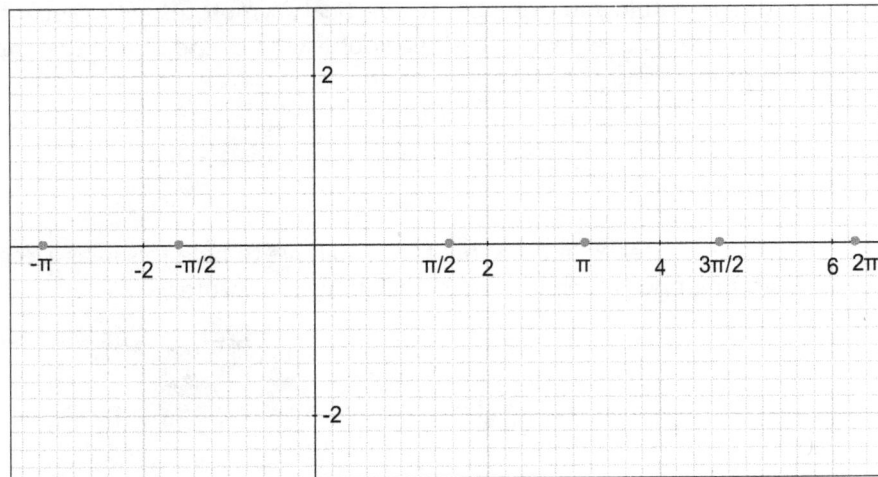

11) Finally, consider the cotangent function. Notice that different forms of the definition have either the sine function or the tangent function in the denominator. Graph both the sine and the tangent functions on GX.

A) What do you notice about places where the two functions equal zero?

B) Where do you expect the vertical asymptotes for the cotangent function to be?

C)   Delete or hide those functions, and graph $y = \cot(x)$, with its asymptotes. You will have to type in $1/(\tan(x))$ or $\cos(x)/\sin(x)$.

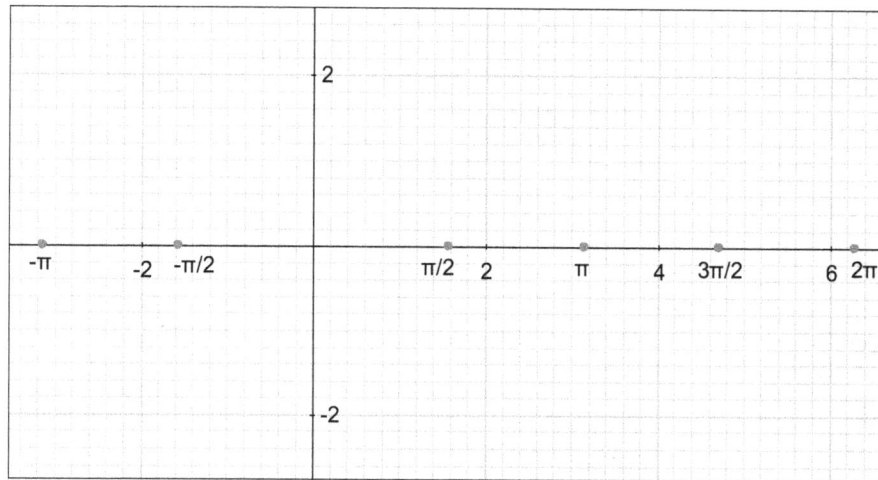

12)   Complete the following statement: "In general, you can expect to find a vertical asymptote in a function graph when ...."